Calcium-Regulating Hormones
I. Role in Disease and Aging

Contributions to Nephrology

Vol. 90

KARGER

Basel · München · Paris · London · New York · New Delhi · Bangkok · Singapore · Tokyo · Sydney

International Symposium on Calcium-Regulating Hormones, Body Functions and Kidney, Nara, Japan, July 11–13, 1990

Calcium-Regulating Hormones

I. Role in Disease and Aging

Volume Editor
H. Morii, Osaka

77 figures and 37 tables, 1991

KARGER

Basel · München · Paris · London · New York · New Delhi · Bangkok · Singapore · Tokyo · Sydney

Cover illustration: This illustration was originally drawn by Dr. Soteki Fuseya [1747–1811), a Japanese scientist of Osaka. In this illustration, an experiment is shown in which perfusion of the renal artery with black carbon sol produced clear fluid (urine) from the ureter, thereby demonstrating the function of blood filtration by the kidney for the first time. Fuseya belonged to the group of Dr. Soukichi Hashimoto (1763–1836) who is regarded as a founder of modern medicine in Osaka. Dr. Kou-An Ogata, who was a famous doctor in Japan in the Edo and Meiji eras, is one of their disciples. Reproduced with permission by Takeda Pharmaceuticals Co., Osaka.

Library of Congress Cataloging-in-Publication Data
International Symposium on Calcium-Regulating Hormones, Body Functions and Kidney (1990: Nara, Japan)
Calcium-regulating hormones / International Symposium on Calcium-Regulating Hormones, Body Functions and Kidney, Nara, Japan, July 11–13, 1990; volume editor H. Morii.
(Contribution to nephrology; vol. 90–91)
Contents: 1. Role in disease and aging – 2. Calcium transport, bone metabolism, and new drugs. Includes indexes.
1. Calcium regulating hormones – congresses. I. Morii, H. (Hirotoshi) II. Title. III. Series.
[DNLM: 1. Bone and Bones – metabolism – congresses. 2. Calcium-metabolism – congresses. 3. Hypertension – Congresses. 4. Kidney failure, Chronic – congresses. 5. Parathyroid Hormones – pharmacology – congresses. 6. Vitamin D – pharmacology – congresses.
W1 CO778UN v. 90–91 / QV 276 I6505c 1990]
QP752.C33I57 1990   599′.01927–dc20
ISBN 3–8055–5451–6 (set)
ISBN 3–8055–5371–4 (v. 90)
ISBN 3–8055–5372–2 (v. 91)

Bibliographic Indices
This publication is listed in bibliographic services, including Current Contents[R] and Index Medicus.

Drug Dosage
The authors and the publisher have exerted every effort to ensure that drug selection and dosage set forth in this text are in accord with current recommendations and practice at the time of publication. However, in view of ongoing research, changes in government regulations, and the constant flow of information relating to drug therapy and drug reactions, the reader is urged to check the package insert for each drug for any change in indications and dosage and for added warnings and precautions. This is particularly important when the recommended agent is a new and/or infrequently employed drug.

© Copyright 1991 by S. Karger AG, P.O. Box, CH-4009 Basel (Switzerland)
Printed in Switzerland on acid-free paper by Thür AG Offsetdruck, Pratteln
ISBN 3–8055–5371–4

Contents

Role of Calcium-Regulating Hormones in the Relationship between Calcium and Hypertension

Role of Endothelin

Calcium and Arteriosclerosis

Calcium and Renal Diseases

Preface

It has become more and more evident that calcium plays an essential role in various aspects of physiological functions and diseases. The International Symposium on Calcium-Regulating Hormones, Body Functions and Kidney held in Japan, July 11–13, 1990, aimed to clarify the problems that exist at the current stage of research. The problems of the relationship between calcium and hypertension, the role of parathyroid hormone in chronic renal failure, and new vitamin D analogs which have limited actions from many effects of $1,25(OH)_2 D_3$, especially of suppressing the secretion of parathyroid hormone, were important topics discussed at the symposium.

The proceedings of the symposium are reported in volumes 90 and 91 of the series 'Contributions to Nephrology'. The papers in this first volume are grouped into 3 sections covering calcium and hypertension, calcium and renal diseases, as well as calcium, diabetes mellitus and aging. McCarron introduces the first section with a discussion of the epidemiological evidence and clinical trials of dietary calcium's effect on blood pressure. The section dealing with calcium and renal diseases is opened by Massry and Fadda's analysis of the mechanism of impaired insulin secretion in chronic renal failure. Fujita's contribution on calcium, parathyroids and aging, which appear to be important factors that interrelate to control the body and cell function, begins the final section.

1. Calcium and Hypertension

Several problems need to be considered in the relationship between calcium and hypertension: (1) calcium and muscle contraction; (2) dietary calcium and hypertension; (3) the role of endothelin, and (4) the mechanism of action of calcium channel blockers.

Calcium and Muscle Contraction

Calcium ion has been shown to be essential in the contraction of actomyosin. Only a few micromolars of calcium ion was needed to activate the contraction of actomyosin [1]. Ebashi [2] first demonstrated the relaxing factor which removes calcium ion from the medium in the presence of ATP and plays an important role in the relaxation of actomyosin. A further contribution by Ebashi [3] was the discovery of troponin which was discovered in the process of studying the discrepancy that calcium was only effective in the crude state of actomyosin and not in pure actomyosin. Troponin has three subunits: inhibitory component L, calcium-binding component C, and tropomyocin-binding component T.

Many factors affecting the intracellular calcium level are related to the contraction of smooth muscle cells in arteries. Matsuura et al. observed that a high salt diet (170 mmol/day) resulted in an increase of calcium ion in platelets, decrease of magnesium in red blood cells, and a decrease in serum ionized calcium levels with a concomitant increase in blood pressure. The intracellular decrease of magnesium may be involved in the increase of ionized calcium resulting in vasoconstriction. Aoki et al. demonstrated that caffeine was less effective in inducing a contraction of the arteries in calcium-depleted sarcoplasmic reticulum in SHR and that the uptake and store of calcium was less in SHR.

Dietary Calcium and Hypertension

McCarron et al. [4] showed in their population study that calcium intake was significantly lower in patients with hypertension. While the increase of cytosolic calcium content is a cause of contraction of actomyosin in smooth muscle cells, how the increase of intracellular calcium occurs in face of the decrease of dietary calcium is a problem. Saito et al. showed in their clinical study that a high salt diet induced a significant increase in blood pressure in the low dietary calcium group in spite of unchanged blood pressure in the high dietary calcium group. They also demonstrated in an experimental study in DOCA-hypertensive rats that the development of hypertension was attenuated by calcium supplementation. Sodium retention was reduced and it may suppress the activity of the sympathetic nervous system in calcium supplementation. Hano et al. concluded from their study in SHR that calcium supplementation normalized calcium handling at the cellular level and suppressed vascular responsiveness.

The intestinal absorption of calcium in hypertensive patients is another problem to be solved. Hennessen et al. reported that the mediated ionized calcium transport into duodenal brush-border membrane vesicles was increased in the adolescent SHR, although the intestinal absorption of calcium in the mature SHR is decreased. However, there are other papers showing contrary results. The authors suspected that the rate-limiting step of perturbed transepithelial calcium transport is localized at the basolateral membrane. Zoccali's paper demonstrated that urinary calcium was increased in patients with essential hypertension at high calcium intake as in stone formers compared to healthy subjects. Serum total calcium fell with low calcium in hypercalciuric hypertensives, but PTH was not different between essential hypertensives and healthy controls. While other authors claimed an elevation of PTH in hypertensives [5], Zoccali et al. did not find a difference in PTH levels between hypertensives and controls, and suggested an inappropriate response of renal calcium excretion to the calcium leak in essential hypertension.

Fujito studied the effect of dietary calcium on sodium ion transport in SHR. When increasing the dietary calcium content, SHR showed a lower blood pressure and lower level of sodium and higher sodium pump activity.

The abnormality in phosphate handling in SHR has been reported. Hirano et al. studied the effects of dietary phosphate intake on phosphate handling in SHR. A lack of adaptation to low phosphate intake was demonstrated in SHR.

Role of Calcium-Regulating Hormones in the Relationship between Calcium and Hypertension

Pang et al. proposed the presence of a hypertensive factor which could be secreted from the parathyroid glands. Reduced blood pressure by parathyroidectomy was restored to the level of hypertension after the transplantation of parathyroid glands in SHR. This factor has been named parathyroid hypertensive factor which is not PTH.

Kishimoto et al., observed the hemodynamic effect of PTHrp in comparison to PTH. The hypotensive effect of PTHrp was more pronounced than that of PTH. Regional hemodynamics were similar between these two hormones but the cardiac output was increased only by PTH.

Jahn et al. performed both in vivo and in vitro studies. In in vivo studies $1,25(OH)_2D_3$ enhanced myocardial contractility, and the sensitivity to catecholamine was increased. However, in in vitro studies using rabbit heart such effects were not proven.

The effect of calcitonin on hemodynamics has well been recognized. However little is known about the chronic effect on blood pressure. Salmon calcitonin has a potency of 1 IU head for 6 days. Blood pressure was measured using the tail-cuff plethysmograhic method. While blood pressure was 180.1 ± 3.0 (SE) mm Hg before and 196.7 ± 3.0 mm Hg after in the vehicle group (n = 18), it was significantly lowered from 184.8 ± 3.0 to 169.4 ± 1.9 mm Hg after the administration of salmon calcitonin

($p < 0.001$, $n = 18$) [6]. Natriuretic and vasodilating actions may contribute to the lowering of blood pressure.

Resnick et al. reported on the effect of salt loading. Significant increases in intracellular pH, calcium and serum $1,25(OH)_2D_3$ and PHF, and decreases in intracellular magnesium occurred in salt-sensitive essential hypertension. They postulated that calcium-regulating hormones may be responsible for salt-sensitive hypertension.

According to the study by Morimoto et al., significant decreases in serum calcium and plasma renin activity, and significant increases in PTH and $1,25(OH)_2D_3$ were observed in senile hypertensive patients.

Whereas calcium-regulating hormones have pharmacological actions on blood pressure, they may also have a pathogenetic significance in essential hypertension. In addition, it is noteworthy that a new hypertensive factor from parathyroid glands was proposed.

Role of Endothelin

Endothelin is one of the most potent vasoconstrictive agents. Takuwa et al. studied the effect of endothelin on aortic smooth muscle cells. There was a biphasic increase in cytosolic calcium. The receptor activation by endothelin is coupled to phospholipase C activation and calcium ion channel gating. Yukimura et al. studied the effect of endothelin on kidney. Renal blood flow, glomerular filtration rate, urine flow and urinary sodium and calcium excretion decreased. The effect on renal blood flow was counteracted by nicardipine but glomerular filtration rate and antinatriuretic and anticalciuretic actions were not.

Koyama et al. observed the secretion of endothelin from porcine aortic endothelial cells. Addition of uremic serum to the incubation medium was followed by a decrease in endothelin secretion. The inhibitory factor was associated with a 50-kD fraction.

Calcium and Arteriosclerosis

Ogihara et al. suggested that the increase in Ca^{2+} mediated by inositol 1,4,5-triphosphate and intracellular alkalinization may function as a signal for enhanced DNA synthesis by LDL in vascular smooth muscle cells.

2. Calcium and Renal Diseases

There are some topics related to vitamin D: metabolism of vitamin D and changes of other parameters as well as treatment of calcium abnormalities in chronic renal diseases.

Vitamin D Metabolism and Other Parameters in Renal Diseases

Mizokuchi et al. reported vitamin D metabolism in the nephrotic rat. Sprague-Dawley rats which were given puromycin aminonucleoside showed lower serum Ca^{2+} and 25(OH)D, and $1,25(OH)_2D_3$ showed a transient increase on the 1st or 2nd day after renal transplantation and eventually normal levels. Nakajima showed transient increase of circulating 1,25(OH)D after renal transplantation. After that the level decreased, and increased gradually again.

The significance of serum bone gla protein has not been established. However, it has recently become possible to measure the intact form of bone gla protein and it is expected that the role of bone gla protein in bone metabolism would be more clearly analyzed. Nakatsuka et al. showed a 20% lower value by a new assay system compared to the previous method. Significant positive correlations were demonstrated between serum bone gla protein and PTH as well as TRACP.

Treatment of Calcium Abnormalities in Chronic Renal Diseases

It is very important to control hypocalcemia which results from a low intake of calcium, hyperphosphatemia and impaired activation of vitamin D. Iseki et al. administered 6.0 g calcium carbonate daily for 6 months to patients with chronic renal insufficiency. There was a significant drop in the urinary excretion of phosphate in the calcium-supplemented group and the occurrence of secondary hyperparathyroidism is retarded. Okamura et al. compared the effects of calcium antagonist and angiotensin-converting enzyme (ACE) inhibitor in patients with chronic renal failure. The ACE inhibitor was more effective in decreasing proteinuria.

Coburn et al. reviewed the clinical application of vitamin D in chronic renal failure. Coen et al. discussed the possibility of application of 1,25-$(OH)_2D_3$ to predialysis patients with chronic renal failure. Goodman and Salusky showed that more than half of the children under peritoneal dialysis demonstrated progressive osteitis fibrosa or failed to respond to treatment with $1,25(OH)_2D_3$.

Nishizawa et al. introduced the use of 26,26,26,27,27,27-hexafluoro-$1,25(OH)_2D_3$ in uremic patients on hemodialysis. It has a higher potency in calcium-mobilizing activity than $1,25(OH)_2D_3$.

3. Calcium, Diabetes mellitus and Aging

Levin et al. [7] reported that loss of skeletal tissue occurs early and is not related to severity, duration or treatment with insulin or diet alone. Shao et al. reported that PTH was elevated in IDDM. Kawagishi et al.

performed the phosphate loading test in patients with NIDDM who responded with a lower level of intact PTH. This discrepancy should be solved in the future, but there could be a difference in the intake of nutrients including calcium, although a decrease in bone mass was observed in DM with a longer duration than 5 years [8].

Although milder hyperparathyroidism in hemodialyzed patients with diabetic nephropathy (DM-HD) has been established [9], Nishitani et al. demonstrated that bone mineral density (BND) was lower in DM-HD and that body weight was an influencing factor for BMD.

Hirotoshi Morii

References

1   Ebashi S: Calcium binding activity of vesicular relaxing factor. J Biochem 1961;50:226–244.
2   Ebashi S: Calcium binding and relaxation in the actomyosin system. J Biochem 1960;48:150–151.
3   Ebashi S, Kodama A: A new protein factor promoting aggregation of tropomyosin. J Biochem 1965;58:107–108.
4   McCarron DA, Morris CD, Henry HJ, Stanton JL: Blood pressure and nutrient intake in the United States. Science 1984;224:1392–1398.
5   McCarron DA, Pigree PA, Rubin RJ, Gauther SM, Molitch M, Krutzik S: Enhanced parathyroid function in essential hypertension: A homeostatic response to a urinary calcium leak. Hypertension 1980;2:162–168.
6   Morii H, Nishizawa Y, Shimazawa E, Takahashi H: Unpublished observation.
7   Levin ME, Boisseau MA, Avioli LA: Effects of diabetes mellitus in juvenile and adult-onset diabetes. N Engl J Med 1976;294:241–245.
8   Morii H, Miki T, Hagiwara S: Unpublished observation.
9   Morii H, Iba K, Nishizawa Y, Miki T, Matsushita Y, Inoue T, Inoue T: Abnormal calcium metabolism in hemodialysed patients with diabetic nephropathy. Nephron 1984;38:24.

Calcium and Hypertension

Morii H (ed): Calcium-Regulating Hormones. I. Role in Disease and Aging.
Contrib Nephrol. Basel, Karger, 1991, vol 90, pp 2–10

Epidemiological Evidence and Clinical Trials of Dietary Calcium's Effect on Blood Pressure

David A. McCarron[1]

Division of Nephrology and Hypertension, Oregon Health Sciences University, Portland, Oreg., USA

The rapid extension of knowledge of the ubiquitous function of calcium and its binding proteins in the regulation of cell function has touched virtually all aspects of human disease [1, 2]. Cardiovascular disorders have been one of the foci of interest. The recognition, clinically, a decade ago that a new category of pharmacologic agents, calcium channel blockers, might represent an effective modality to lower blood pressure in humans, heightened awareness of a potentially specific role for altered cellular Ca^{2+} homeostasis in the pathogenesis of human hypertension.

Data from metabolic assessments of humans with essential hypertension [3] and a commonly studied experimental model, the spontaneously hypertensive rat (SHR) [4], have also indicated that the state of calcium homeostasis may be a predictor of the blood pressure level. The hypothesis, while very controversial [5], that inadequate dietary calcium intake might be a previously unrecognized factor in the pathogenesis of increased arterial pressure was proposed [6, 7]. As data and experience in this area accumulates, the potential application of maintenance of some, as yet undefined, optimal level of dietary calcium to reduce either the prevalence of hypertension in the population or the level of blood pressure of an individual has been recognized by a variety of authorities [8, 9]. We have the opportunity by either increasing calcium intake or prescribing calcium channel blockers to selected patients, to produce beneficial cardiovascular responses. Only dietary intake, though, holds the potential for application as a preventive intervention to achieve that end in the general population.

The scientific basis for that position has recently been summarized [10]. Based primarily on the epidemiological reports of dietary calcium's

[1] I wish to thank Molly Reusser for her editorial assistance and John Davis for preparing the manuscript. Much of the author's original research cited in this publication has been supported by a grant from the National Dairy Promotion and Research Board, administered through the USDA.

effects on blood pressure and, to a lesser extent, observations from the laboratory, the notion that a threshold exists for the particular nutrient/ blood pressure relationship will be developed in the present paper. The impact of demographic, lifestyle, nutritional and clinical conditions on the 'set' of this threshold will be assessed. This approach will be used to develop the hypothesis that the increased prevalence of hypertension in selected segments of our society may well reflect variations in the 'set' of the threshold. To the degree that clinical trials are poorly designed it will be further argued that the variable outcome of the trials of calcium supplementation to treat high blood pressure would be anticipated as the population studied and the protocols used have in many cases been inadequate. A thoughtful assessment of the clinical trials reported to date can account for much of the variation in outcome observed and does not negate the potential importance of encouraging all adults to maintain a dietary calcium intake at or above 800–1,200 mg/day for purposes of reducing their risk of developing high blood pressure.

Epidemiology

Our publication, in 1982, of the epidemiological association between reported dietary calcium intake and blood pressure status [6] has been followed by at least 24 reports in the medical literature. Virtually all have identified this association. A recent review summarized those reports [10]. One of the more recent publications in this field, that of Witteman et al. [9], provides the most compelling evidence to date. They identified a reduction in the risk of developing hypertension from a prospective longitudinal 4-year survey [9]. All the other reports have been based upon the cross-sectional analysis of a population database.

The association of lower blood pressures with higher intakes of dietary calcium has been observed in population surveys as well as in more limited samples. It has been observed in a variety of ethnic groups (Japanese, Italians, Dutch, Puerto Ricans, Russians and Chinese), in males and females, in age groups that span infancy to the eighth decade of life, in different racial groups and in specific settings associated with increased blood pressure risk (excessive alcohol intake, pregnancy, and high sodium chloride intake). The strength of the association, while modified by other variables known to predict blood pressure, such as alcohol and BMI, has been remarkably consistent as an independent predictor of blood pressure status.

The relation between blood pressure and dietary calcium intake has been linked to other nutritional factors. These have included potassium

[11], magnesium [9], phosphorus [7, 9], sodium chloride [12] and alcohol [13]. However, adjustments for these other dietary constituents have typically not eliminated the association. Dairy products have been identified as the source of the dietary calcium linked with blood pressure status in most, but not all, of the reports [6, 7, 9, 11, 14].

With the analysis of the HANES I database we reported in 1984, an apparent threshold for the effect of increasing dietary calcium on reducing blood pressure was first suggested [7]. Though our publication did not specifically comment on this characteristic of the relationship, it was evident. We based our assessment of a threshold effect upon the observation that as the level of dietary calcium in the US population was lowered from 1,400 to 600 mg/day, the proportion of subjects with elevated blood pressure rose from 4–5 to 7–8%; however, with the next 300 mg/day reduction (600–300 mg/day) the proportion of Americans with elevated blood pressure increased an additional 3–4%. Thus, the inverse relationship is not strictly linear. The slope is greater below 600 mg/day than it is above this level of daily calcium consumption.

We further observed in analyzing the dietary calcium relationship that the threshold varied on the basis of gender and age. For example, younger women (<55 years) demonstrated a virtually flat relationship down to 300 mg/day, where males from the same age range exhibited a threshold at 700–800 mg/day. Compared to women less than 55 years of age, the older females in the HANES I database appeared to have a threshold for the protective effect of dietary calcium that was in the range of 700 mg/day.

Subsequent reports from other databases [9, 13, 15] as well as the HANES I survey [12], have also identified a threshold effect. The second report from the Honolulu Heart Project specifically addressed this issue [15]. For the middle-aged, Japanese-American males in that study, the threshold was estimated to lie below 300 mg/day, a level of dietary calcium consumed by 25% of the subjects in that study population.

In their analytical approach to the HANES I data, Gruchow et al. [12] concluded that only below a specific level of dietary calcium (400 mg) was there evidence that higher levels of dietary sodium chloride and lower levels of potassium were associated with increased arterial pressure [12]. Besides confirming that a calcium threshold existed in the HANES I data, these authors' findings raised the possibility that the level of at least two other dietary electrolytes (Na$^+$, K$^+$) might be a factor in determining the 'set' of this calcium threshold.

Criqui et al. [13], in their most recent report from the Honolulu Heart Project, noted that the quantity of alcohol consumed by the males in this study was a determining factor in whether a blood-pressure-lowering effect of dietary calcium could be identified. In the males who were defined as moderate-to-heavy drinkers (>13.3 ml of alcohol per day), it was not possible to show any protective effect of the dietary calcium on blood pressure. The authors concluded that some effect of alcohol, likely an impairment of dietary calcium absorption, produced a calcium-depleted state [13]. Even though reported calcium intakes in the 'moderate-plus' subgroup were at levels in many of the subjects that would otherwise be expected to be associated with lower blood pressures, they were not afforded any protection by the dietary calcium.

A reasonable and logical extension of the Criqui et al. [13] findings is that simply ingesting what might be considered an adequate level of dietary calcium does not ensure that a blood pressure protective effect will be observed. Factors which might impair absorption, as they speculated alcohol had done in these Japanese-American males in their report, or enhance excretion would have the effect of shifting the threshold to much higher levels at which dietary calcium's cardioprotective action may be demonstrable. They also noted the reverse phenomenon, an enhancement of the calcium effect at higher levels of dietary potassium.

The Nurses' Health Study has recently reaffirmed the hypothesis that the relation between dietary calcium intake and blood pressure is not strictly inverse and linear [9]. Witteman et al. [9] identified a level of 600 mg/day or more of calcium as necessary to show a meaningful reduction in the risk of developing hypertension over a 4-year period. A minimum of 400–500 mg/day of the dietary calcium of these women aged 34–59 was derived from dairy sources. In those nurses who derived the majority of their dietary calcium from nondairy sources, comparable threshold effects of the calcium are observed. An additional confirmatory finding was that the threshold was shifted to significantly higher levels of calcium in those women who consumed more than 20 g of alcohol each day. As was observed in the Honolulu data, this impact of alcohol on calcium metabolism and its cardiovascular actions was unique' for that cation as a similar effect was not reported for magnesium or other nutrients.

Intervention Trials

Over 20 clinical trials testing the hypothesis that by increasing calcium intake in humans blood pressure will be decreased have been reported. Cutler and Brittain [16], as well as Grobbee and Wall-Manning [17] and Mikami et al. [18] have recently reviewed the results of many of these trials from the fromer two groups using a meta-analysis approach. All three analyses concluded that overall a beneficial effect, i.e. reduction in blood pressure, was observed. However, the impact was only significant for systolic blood pressure, though a trend was evident for diastolic pressure also.

Grobbee and Wall-Manning [17] specifically addressed the issue of why, with such a consistent protective effect apparent in the epidemiological data, a stronger effect was not demonstrable from human trials. He acknowledged the well-known limitations of the meta-analysis approach, a weakness largely ignored by Cutler and Brittain [16] in their discussion. Principally, the weaknesses surround, comparing in the same analysis, studies which varied so widely in their design and the end-points employed. While clearly useful, the meta-analysis must be interpreted in the light of known factors that could significantly influence the outcome.

The most important factor at the outset is the inclusion in the analysis of all published trials and the exclusion of reports based upon abstracts and preliminary communications where insufficient data are presented. The importance of this approach is highlighted by the striking differences between Cutler and Brittain's [16] approach and that of Mikami et al. [18] in terms of the studies included and the categorization of whether a trial did or did not show a reduction. The latter were more thorough in their survey with the inclusion of studies by Luft et al. [19] and Tabuchi et al. [20] and which were excluded by Cutler and Brittain [16]. Since the Luft et al. [19] study included a placebo control, that factor cannot explain the exclusion by these authors. Importantly, all of those studies not cited by Cutler and Brittain [16] were ones that had demonstrated a significant effect of increasing calcium intake on lowering blood pressure.

Even if an author includes all the appropiate studies, a significant bias can still be introduced in the meta-analysis depending upon how the endpoint outcome is defined. As an example, in the Cutler-Brittain analysis, the trial by Strazzullo et al. [21] was categorized as having shown no effect on blood pressure, and yet, as cited by Mikami et al. [18], this trial reported a significant reduction in blood pressure, exhibiting a time-course and magnitude of impact remarkably similar to that reported by us in an earlier trial in a similar, though larger population [22]. Likewise, the trial by Lasaridis et al. [23] was categorized differently by these two analyses [16, 18]. Since it is unlikely that a report would be categorized as showing a significant effect when one was not observed, interpretive bias by the authors of this meta-analysis must have accounted for this disparity.

Based upon the information available from the epidemiological studies several criteria emerge as possibly important predictors of a positive outcome in such an intervention trial. Many have been noted above. They include gender, as it appears that younger women exhibit little sensitivity to calcium's antihypertensive action. Thus, it would be anticipated that a trial involving young women would not demonstrate any effect and, indeed, that was the experience of Thomsen et al. [24] and Van Beresteyn et al. [25]. Other dietary factors must also be taken into consideration. Since a concurrent higher sodium intake appears to influence favorably the blood pressure response to increasing calcium intake [26], it is not surprising that Nowson and Morgan [27] did not observe a significant effect as their subjects were all on a sodium-chloride-restricted diet. In contrast, the study of Saito et al. [28] brought out this important interaction and consideration in study design, as an increase in dietary calcium intake blocked any pressor response to a 300-mEq NaCl diet in their population of hypertensive adults. The human trials reported by Resnick et al. [29, 30] had previously documented this important dependence of sodium chloride and calcium on their respective blood pressure effects. Any calcium intervention trial that purposely or inadvertently modifies one of the other nutrients (see above) now believed to influence the action of dietary calcium on blood pressure needs to acknowledge these additional confounding variables and their potential input on the outcome considered.

Five additional logistical considerations must be assessed when evaluating the blood pressure response to calcium intervention trials. First, the sample size used must be considered. Using a cross-over, placebo-controlled design where blood pressures are measured weekly, at least 40 subjects must be included. Where a parallel design is used, each group probably needs to include 40 individuals or 80 total. As summarized in both the articles of Mikami et al. [18] and Cutler and Brittain [16], virtually all the 'negative' trials done in hypertensive adults failed to satisfy these sample size requirements [27, 31–36]. Second, because of the inherent variability of blood pressure measurements, an adequate number of blood pressure readings must be incorporated. A number of the reported trials based their conclusions on only one or two blood pressures after the initiation of calcium intervention [31, 33, 34]. Increasing the frequency and intensity of the blood pressure measurements will readily compensate for a smaller sample size. The studies of Luft et al. [19], Tabuchi et al. [20] and Saito et al. [28] were all carried out under 'clinical research center' conditions, i.e. daily or more frequent blood pressure measurements.

Third, in four of the earliest trials [22, 37–39], a time dependency of the onset of the blood pressure reduction was identified. Using weekly or bi-weekly blood pressure measurements, a minimum of 4–6 weeks were required before a blood pressure effect was observed. As with the sample size criterion, the duration of the intervention can be truncated if very frequent (daily) measurements of blood pressure are incorporated into the study design. In fact, as demonstrated by Luft et al. [19], Tabuchi et al. [20], and Saito et al. [28], a reduction in blood pressure can be observed within 1–2 weeks if blood pressure is monitored frequently. Cutler and Brittain [16] were unable to identify a duration effect, likely because of the presence in these trials of short length, but frequent monitoring. Fourth, a proper control must be incorporated in the protocol. Unfortunately in several of the reported studies the control was difficult to identify [27, 31, 34], an error that is typically compounded by too few blood pressure measurements and too limited a duration of the intervention. Fifth, Cutler and Brittain [16] did note a dose effect. Based upon the discussion above, it is unlikely that trials that employed less than 1,000 mg would have demonstrated an effect and, indeed, they did not [27, 32, 33, 39].

Conclusion

In this review I have summarized the strengths and variability of the database associating the level of dietary calcium to the prevalence of hypertension in humans. The notion that this relationship is characterized by the threshold, below which the cardiovascular risk increases at a greater rate, is suggested by several of the epidemiological reports. The likelihood that this threshold's 'set-point' can be modified by demographic, nutritional, lifestyle and genetic factors is evident based upon a variety of reported findings. Finally, I would propose that the specific application of increasing dietary calcium intake, within the already recommended range for this nutrient, holds the greatest potential in a number of specific clinical settings (Blacks, Asians (Japanese), alcoholics, diabetics, salt-sensitive subjects, pregnant women, elderly individuals). In each, the risk of hypertension, compared to the general population, is significantly greater and either dietary calcium intake is low or a supervening factor imposes a higher level of dietary calcium to maintain optimal homeostasis.

Summary

The intervention trials intended to assess prospectively the efficacy of increasing calcium intake to lower blood pressure have in general been inconclusive in their results. If, however, they are evaluated in terms of what had been suggested by the epidemiological surveys as to the conditions that would have to be met, then a more consistent pattern emerges. When one requires a combination of either an adequate sample size with frequent enough blood pressure determinations carried out in a population projected to be sensitive to this life-style (diet) intervention, then the results are remarkably consistent. Excluding the trials of short duration and/or limited blood pressure determinations plus those carried out in young females, a consistent benefit from increasing calcium intake for reducing arterial pressure is apparent in both normotensive and hypertensive males and females.

References

1   Rasmussen H, Barrett PQ: Calcium messenger system: An integrated view. Physiol Rev
     1984;64:938–984.
2   Cheung WY: Calmodulin plays a pivotal role in cellular regulation. Science
     1980;207:19–27.
3   McCarron DA, Pingree PA, Rubin RJ, Gaucher SM, Molitch M, Krutzik S: Enhanced
     parathyroid function in essential hypertension: A homeostasis response to a urinary
     calcium leak. Hypertension 1980;2:162–168.
4   McCarron DA, Yung NN, Ugoretz BA, Krutzik S: Disturbance of calcium metabolism
     in the spontaneously hypertensive rat (SHR): Attenuation of hypertension by calcium
     supplementation. Hypertension 1981;3(suppl 1):162–167.
5   Kaplan NM, Meese RB: The calcium deficiency hypotension of hypertension: A
     critique. Ann Intern Med 1986;105:947–955.
6   McCarron DA, Morris CD, Cole C: Dietary calcium in human hypertension. Science
     1982;217:267–269.
7   McCarron DA, Morris CD, Henry HJ, Stanton JL: Blood pressure and nutrient intake
     in the United States. Science 1984;224:1392–1398.
8   Hypertension: Is there a place for calcium (Editorial)? Lancet 1986;i:359–361.
9   Witteman JCM, Willett WC, Stampter MJ, et al: A prospective study of nutritional
     factors and hypertension among US women. Circulation 1989;80:1320–1327.
10   McCarron DA: Calcium metabolism and hypertension. Kidney Int 1989;35:717–736.
11   Reed D, McGee D, Yamo K, Hankin J: Diet, blood pressure and multicollinearity.
     Hypertension 1985;7:405–411.
12   Gruchow HW, Sobocinski KA, Barboriak JJ: Calcium intake and the relationship of
     dietary sodium and potassium to blood pressure. Am J Clin Nutr 1988;48:1463–1470.
13   Criqui MH, Langer RD, Reed DM: Dietary alcohol, calcium, and potassium: Indepen-
     dent and combined effects on blood pressure. Circulation 1989;80:609–614.
14   Ackley S, Barrett-Connor E, Suarez L: Dairy products, calcium and blood pressure. Am
     J Clin Nutr 1983;38:457–461.
15   Joffres MR, Reed DM, Yano K: Relationship of magnesium intake and dietary factors
     to blood pressure: The Honolulu Heart Study. Am J Clin Nutr 1987;45:469–475.
16   Cutler JA, Brittain E: Calcium and blood pressure: An epidemiologic perspective. Am J
     Hypertens 1990;3:137S–146S.
17   Grobbee DE, Wall-Manning HJ: The role of calcium supplementation in the treatment
     of hypertension. Drugs 1990;39:7–18.
18   Mikami H, Ogihara T, Tabuchi Y: Blood pressure response to dietary calcium interven-
     tion in humans. Am J Hypertens 1990;3:137S–146S.
19   Luft FC, Aronoff GR, Sloan RS, et al: Short-term augmented calcium intake has no
     effect on sodium homeostasis. Clin Pharmacol Ther 1986;39:414–419.
20   Tabuchi Y, Ogihara T, Hashizume K, et al: Hypotensive effect of long-term oral calcium
     supplementation in elderly patients with essential hypertension. J Clin Hypertens
     1986;245–262.
21   Strzzullo P, Siani A, Guglielmi S, et al: Controlled trial of long-term oral calcium
     supplementation in essential hypertension. Hypertension 1986;8:1084–1088.
22   McCarron DA, Morris CD: Blood pressure response to oral calcium in persons with
     mild to moderate hypertension: A randomized, double-blind, placebo-controlled,
     crossover trial. Ann Intern Med 1985;103:825–831.
23   Lasaridis AN, Kaisis CN, Zanairi KI, et al: Oral calcium supplementation promotes renal
     sodium excretion in essential hypertension. J Hypertens 1987;5(suppl 5):S307–S309.

24 Thomsen K, Nilas L, Christensen C: Dietary calcium intake and blood pressure in normotensive subjects. Acta Med Scand 1987;222:51–56.

25 Van Beresteyn ECH, Schaafsma G, de Waard H: Oral calcium and blood pressure: A controlled intervention trial. Am J Clin Nutr 1986;44:883–888.

26 McCarron DA, Lucas PA, Shneidman RS, Drüeke T: Blood pressure development of the spontaneously hypertensive rat following concurrent manipulation of the dietary Ca^{2+} and Na^{+}: Relation to intestinal Ca^{2+} fluxes. J Clin Invest 1985;76:1147–1154.

27 Nowson C, Morgan T: Effect of calcium carbonate on blood pressure. J Hypertens 1986;4(suppl 6):S673–S675.

28 Saito K, Sano F, Furuta Y, et al: Effect of oral calcium on blood pressure response in salt-loaded borderline hypertensive patients. Hypertension 1989;13:219–226.

29 Resnick LM, Nicholson JP, Laragh JH: Outpatient therapy of essential hypertension with dietary calcium supplementation. J Am Coll Cardiol 1984;3:616.

30 Resnick LM, Difabio B, Marion RM, et al: Dietary calcium modifies the pressor effects of dietary salt intake in essential hypertension. J Hypertens 1986;4(suppl 6):S679–S681.

31 Sunderrajan B, Bauer JH: Oral calcium supplementation does not alter blood pressure or vascular response in normotensive men. Circulation 1984;70:II–130.

32 Meese RB, Gonzales DG, Casparian JM, et al: The inconsistent effects of calcium supplements upon blood pressure in primary hypertension. Am J Med Sci 1987;249:219–224.

33 Cappuccio FP, Markandu ND, Singer DR: Does oral calcium supplementation lower blood pressure? A double blind study. J Hypertens 1987;5:67–71.

34 Zoccali C, Mallamaci F, Delfino D, et al: Does calcium have a dual effect on arterial pressure? J Hypertens 1987;5(suppl 5):S267–S269.

35 Zoccali C, Mallamaci F, Delfino D: Double-blind randomized, crossover trial of calcium supplementation in essential hypertension. J Hypertens 1988;6:451–455.

36 Siani A, Strazzullo P, Guglielmi S: Controlled trial of low calcium versus high calcium intake in mild hypertension. J Hypertens 1988;6:253–256.

37 Belizan JM, Villar J, Pineda O, et al: Reduction of blood pressure with calcium supplementation in young adults. JAMA 1983;249:1161–1165.

38 Lyle RM, Melby CL, Hyner GC, et al: Blood pressure and metabolic effects of calcium supplementation in normotensive white and black men. JAMA 1987;257:1772–1776.

39 Grobbee DE, Hoffman A: Effect of calcium supplementation on diastolic blood pressure in young people with mild hypertension. Lancet 1986;ii:703–707.

40 Vinson JA, Mazur T, Bose P: Comparison of different forms of calcium on blood pressure of normotensive young males. Nutr Rep Int 1987;36:497–505.

David A. McCarron, MD, Department of Medicine, Division of Nephrology,
Oregon Health Sciences University, 3181 S.W. Sam Jackson Park Road,
Portland, OR 97201 (USA)

Morii H (ed): Calcium-Regulating Hormones. I. Role in Disease and Aging.
Contrib Nephrol. Basel, Karger, 1991, vol 90, pp 11–18

Significance of Intracellular Cations and Calcium-Regulating Hormones on Salt Sensitivity in Patients with Essential Hypertension

Hideo Matsuura[a], *Tetsuji Shingu*[a], *Ichiro Inoue*[a], *Goro Kajiyama*[a],
Miho Kusaka[b], *Koji Matsumoto*[b], *Koji Kido*[c]

[a] 1st Department of Internal Medicine, Hiroshima University School of Medicine;
[b] Division of Internal Medicine, Saiseikai Hiroshima Hospital, Japan; [c] 2nd Department
of Internal Medicine, Hiroshima Hospital of West Japan Railway Company, Hiroshima,
Japan

It is well known that excessive intake of sodium chloride (NaCl) results in high blood pressure. As the possible mechanisms of pressor response to oral NaCl loading, volume retention [1], exaggerated sympathetic nerve activity [2], increased response to pressor substances [3], and abnormalities in intracellular cations metabolism [4, 5] have been reported.

de Wardener and MacGregor [6] and Blaustein [7] proposed the hypothesis in regard to the etiology of essential hypertension. This hypothesis is convenient for explaining the pressor response implicated in NaCl loading, i.e. salt sensitivity, as we previously reported using erythrocytes and lymphocytes [8]. In addition to the intracellular cations metabolism, extracellular cations metabolism, especially calcium and magnesium, and calcium-regulating hormones seem to be important, because the negative balance of calcium [9] or magnesium [10] has been reported in essential hypertension.

In this issue, we focused our interests on the relation between intra- and extracellular cations, calcium-regulating hormones and blood pressure in the essential hypertensives with salt loading.

Patients and Methods

Eighteen inpatients with mild-to-moderate essential hypertension (10 male and 8 female, 48.7 ± 12.9 years old) were studied. Informed consent was obtained from all patients. Antihypertensive drug therapy was stopped at least 4 weeks prior to this study. After admission, regular salt diet (170 mmol NaCl/day) was served for 1 week followed consecutively by a low salt diet (50 mmol NaCl/day) and a high salt diet (340 mmol NaCl/day) for 1 week. During the study period, the intakes of potassium (50 mmol/day), calcium (15 mmol/day) and total calories (40 cal/kg/day) were kept constant.

On the morning of the last day of each diet period, the resting blood pressure was measured under fasting conditions in the supine position. Blood was drawn through an indwelling catheter for the determination of serum electrolyte concentrations (Na, K, Ca, Mg), plasma renin activity (PRA), parathyroid hormone (PTH) concentrations, and calcitonin (CT), serum-ionized calcium concentration ($[Ca^{2+}]_o$), intraerythrocyte magnesium concentration ($E[Mg]_i$), and intraplatelet free calcium concentration ($PLT[Ca^{2+}]_i$). Urinary electrolyte excretions were also determined.

PRA, PTH and CT were measured by radioimmunoassay. $[Ca^{2+}]_o$ was determined by the ion-electrode method [11]. $E[Mg]_i$ was measured by an atomic absorption spectrophotometer using the hematocrit capillary method [12]. $PLT[Ca^{2+}]_i$ was determined using Fura-2 according to the method of Pollock et al. [13].

The data are shown as means $\pm SD$. Comparison of variables between the low and high salt periods was performed by Wilcoxon's signed-rank test. Correlation was analyzed using the Spearman's rank-order correlation. Probability values < 0.05 were considered to be significant.

Results

The variables during the low and high salt diets are shown in table 1. After oral NaCl loading, mean blood pressure was increased. Urinary excretion of sodium was increased and PRA was suppressed. Urinary

Table 1. Variables during low salt diet (50 mmol NaCl/day) and high salt diet (340 mmol NaCl/day) in patients with essential hypertension

	Low salt	High salt
Mean blood pressure, mm Hg	107.4 ± 15.5	$114.5 \pm 12.7^*$
Heart rate, beats/min	65.2 ± 12.8	63.8 ± 14.0
Urinary sodium excretion, mmol/day	46.4 ± 15.8	$228.0 \pm 57.3^{**}$
Urinary potassium excretion, mmol/day	33.2 ± 9.2	36.7 ± 10.7
Urinary calcium excretion, mmol/day	5.89 ± 2.25	$8.63 \pm 3.56^{**}$
Urinary magnesium excretion, mmol/day	66.6 ± 15.8	$74.0 \pm 19.6^{**}$
Serum sodium concentration, mmol/l	143.8 ± 1.7	144.9 ± 2.3
Serum potassium concentration, mmol/l	4.28 ± 0.36	$3.92 \pm 0.31^{**}$
Serum calcium concentration, mmol/l	2.35 ± 0.11	$2.25 \pm 0.10^{**}$
Serum magnesium concentration, mmol/l	1.10 ± 0.24	$1.00 \pm 0.28^{**}$
Serum ionized calcium concentration, mmol/l	1.13 ± 0.03	$1.11 \pm 0.04^{**}$
Plasma renin activity, ng angiotensin I/ml/h	1.91 ± 1.49	$0.46 \pm 0.40^{**}$
Plasma parathyroid hormone concentration, pg/ml	311.1 ± 91.5	308.9 ± 104.1
Plasma calcitonin concentration, pg/ml	36.7 ± 13.6	34.1 ± 12.3
$[Mg]_i$ in erythrocytes, mmol/l cells	2.88 ± 0.17	$2.77 \pm 0.20^{**}$
$[Ca^{2+}]_i$ in platelets, nmol/l	93.6 ± 35.0	$120.5 \pm 31.6^{**}$

* p < 0.05, ** p < 0.01, compared with values during low salt diet.

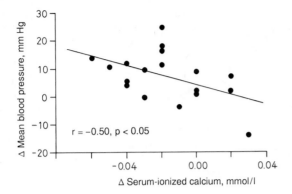

Fig. 1. Relation between changes (Δ) in mean blood pressure and serum-ionized calcium concentration with NaCl loading in patients with essential hypertension.

calcium and magnesium excretions were increased and serum potassium, calcium and magnesium concentrations were decreased slightly but significantly. $[Ca^{2+}]_o$ also showed a significant decrease. $E[Mg]_i$ was decreased and $PLT[Ca^{2+}]_i$ was increased significantly.

The changes in (Δ) mean blood pressure by NaCl loading showed an inverse relationship with $\Delta[Ca^{2+}]_o$ ($r = -0.50$, $p < 0.05$; fig 1), whereas Δmean blood pressure did not correlate with the changes in urinary electrolyte excretion and in serum electrolyte concentrations. ΔMean blood pressure correlated with ΔPTH positively ($r = 0.49$, $p < 0.05$) but did not correlate with ΔCT and ΔPRA. ΔMean blood pressure correlated inversely with $\Delta E[Mg]_i$ ($r = -0.56$, $p < 0.05$; fig. 2a) and positively with $\Delta PLT[Ca^{2+}]_i$ ($r = 0.46$, $p < 0.05$; fig. 2b). $\Delta PLT[Ca^{2+}]_i$ correlated inversely with ΔPRA ($r = -0.53$, $p < 0.05$), $\Delta[Ca^{2+}]_o$ ($r = -0.66$, $p < 0.01$; fig. 3a) and $\Delta E[Mg]_i$ ($r = -0.49$, $p < 0.05$; fig. 3b) and positively with ΔPTH ($r = 0.44$, $p = 0.05$).

Discussion

We have reported that oral NaCl loading induces the increases in intracellular sodium and free calcium concentrations and that the increases in these cations correlate with the changes in mean blood pressure by NaCl loading, i.e. salt sensitivity [8, 14]. These results are compatible with the sodium transport hypothesis [6, 7]. In this study, $[Ca^{2+}]_i$ was determined using platelets, since platelets are supposed to be a suitable model for

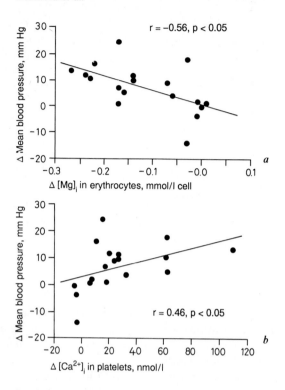

Fig. 2. Relation between changes (Δ) in mean blood pressure and [Mg]$_i$ in erythrocytes (*a*) and [Ca^{2+}]$_i$ in platelets (*b*) with NaCl loading in patients with essential hypertension.

vascular smooth muscle cells [15]. ΔPLT[Ca^{2+}]$_i$ showed a positive correlation with Δmean blood pressure as we demonstrated previously using lymphocytes [14].

NaCl loading enhanced the renal calcium and magnesium excretions and decreased the serum levels of these divalent cations. It is possible that decreased [Ca^{2+}]$_o$ elevates [Ca^{2+}]$_i$ by itself or through changes in calcium-regulating hormones, since ΔPLT[Ca^{2+}] inversely correlated with Δ[Ca^{2+}]$_o$ and positively with ΔPTH. Although there is a positive correlation between ΔPLT[Ca^{2+}]$_i$ and ΔPTH, it is not clear whether PTH directly contributes to the changes in [Ca^{2+}]$_i$ because PTH induces relaxation of vascular smooth muscle of spontaneous hypertensive rats in vitro [16], as opposed to hyperparathyroidism in which hypertension has been observed [17]. Since it was reported that 1,25(OH)$_2$D$_3$ increased [Ca^{2+}]$_i$ of vascular

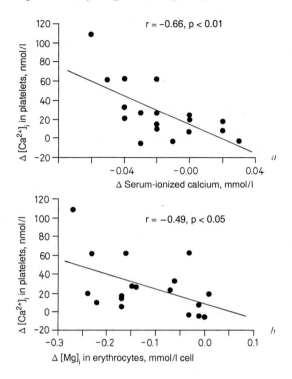

Fig. 3. Relation between changes (Δ) in [Ca^{2+}]$_i$ in platelets and serum-ionized calcium concentration (*a*) and [Mg]$_i$ in erythrocytes (*b*) with NaCl loading in patients with essential hypertension.

smooth muscle cells [18, 19], the elevated [Ca^{2+}]$_i$ as shown in this study may be explained by increased 1,25(OH)$_2$D$_3$ due to PTH stimulation. However, more precise studies are required to clarify the roles of these calcium-regulating hormones on [Ca^{2+}]$_i$ metabolism and salt sensitivity.

Since magnesium is a physiological calcium blocker [20] and decreases [Ca^{2+}]$_i$ by stimulating Na-K ATPase [13, 21] and Ca ATPase in cell membrane [22] or sarcoplasmic reticulum [23], the decreases in [Mg]$_o$ and [Mg]$_i$ are possible to increase [Ca^{2+}]$_i$ through the increase in calcium influx or the inactivation of these ATPases. These possibilities were supported by the present results that serum magnesium was decreased by NaCl loading and that ΔE[Mg]$_i$ correlated with ΔPLT[Ca^{2+}]$_i$ inversely.

Our previous results and these suggest a relationship between salt sensitivity, cation metabolism and calcium-regulating hormones (fig. 4).

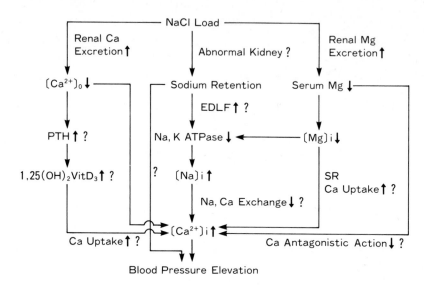

Fig. 4. Possible roles of cations metabolism and calcium-regulating hormones in relation to salt sensitivity in patients with essential hypertension. EDLF = Endogenous digitalis-like factor.

In conclusion, the blood pressure response to oral NaCl loading is associated with the alteration of intracellular calcium metabolism. The magnesium metabolism and calcium-regulating hormones may be involved in the increase of $[Ca^{2+}]_i$ resulting in vasoconstriction.

Summary

Although the existence of salt sensitivity in essential hypertensives has been well known, the precise mechanism(s) has not yet been elucidated. The aim of this study was to clarify the relation between the responses in blood pressure, extra- and intracellular cations and calcium-regulating hormones to oral NaCl loading in essential hypertensives. After oral NaCl loading, mean blood pressure, urinary excretions of calcium and magnesium, and $PLT[Ca^{2+}]_i$ were significantly increased. $[Ca^{2+}]_o$ and $E[Mg]_i$ were decreased. The changes (Δ) in mean blood pressure by NaCl loading positively correlated with $\Delta PLT[Ca^{2+}]_i$ and ΔPTH, and negatively with $\Delta[Ca^{2+}]_o$ and $\Delta E[Mg]_i$. $\Delta PLT[Ca^{2+}]_i$ positively correlated with ΔPTH and negatively with $\Delta[Ca^{2+}]_o$ and $\Delta E[Mg]_i$.

From these results, the blood pressure response to oral NaCl loading is associated with the alteration of $[Ca^{2+}]_i$ metabolism in which the changes in magnesium metabolism and calcium-regulating hormones may be involved.

References

1 Kawasaki T, Hatano K, Weiss GB: The effect of high-sodium and low-sodium intakes on blood pressure and other related variables in human subjects with idiopathic hypertension. Am J Med 1978;64:193–198.

2 Fujita T, Henry WL, Bartter FC, Lake CR, Delea CS: Factors influencing blood pressure in salt-sensitive patients with hypertension. Am J Med 1980;69:334–344.

3 Rankin LI, Luft FC, Henry DP, Gibbs PS, Weinberger MH: Sodium intake alters the blood pressure effect of norepinephrine. Hypertension 1981;3:650–656.

4 Haddy FJ: Abnormalities of membrane transport in hypertension. Hypertension 1983;5(suppl V):V66–V72.

5 Postnov YV, Orlov SN, Shevchenko A, Adler AM: Altered sodium permeability, calcium binding and Na-K-ATPase activity in the red blood cell membrane in essential hypertension. Pflügers Arch 1977;371:263–269.

6 de Wardener HE, MacGreger GA: Dahl's hypothesis that a saluretic substance may be responsible for a sustained rise in arterial pressure: Its possible role in essential hypertension. Kidney Int 1980;18:1–9.

7 Blaustein MP: Sodium ions, calcium ions, blood pressure regulation, and hypertension: A reassessment and a hypothesis. Am J Physiol 1977;232:C165–C173.

8 Kido K, Matsuura H, Otsuki T, Matsumoto K, Shingu T, Oshima T, Inoue I, Kajiyama G: Sodium chloride sensitivity, intracellular sodium concentration in erythrocytes and lymphocytes, and renin profile in essential hypertension. Jpn Circ J 1989;53:101–107.

9 MaCarron DA: Calcium metabolism and hypertension. Kidney Int 1989;35:717–736.

10 Altura BM, Altura BT, Gebrewold A, Ising H, Gunther T: Magnesium deficiency and hypertension: Correlation between magnesium-deficient diet and microcirculatory changes in situ. Science 1984;223:1315–1317.

11 Shore AC, Booker J, Sagnella GA, Markandu ND, MacGregor GA: Serum ionized calcium and pH: Effects of blood storage, some physiological influences and a comparison between normotensive and hypertensive subjects. J Hypertens 1987;5:499–505.

12 Oshima T, Matsuura H, Kido K, Matsumoto K, Otsuki T, Shingu T, Inoue I, Kajiyama G: Intracellular sodium and potassium concentrations in erythrocytes of healthy male subjects. Jpn J Nephrol 1988;30:1095–1101.

13 Pollock WK, Rink TJ, Irvine RF: Liberation of [^{3}H]-arachidonic acid and changes in cytosolic free calcium in fura-2 loaded human platelets stimulated by ionomycin and collagen. Biochem J 1986;235:869–877.

14 Oshima T, Matsuura H, Kido K, Matsumoto K, Otsuki T, Fujii H, Masaoka S, Okamoto M, Tsuchioka Y, Kajiyama G, Tubokura T: Intralymphocytic sodium and free calcium concentration in relation to salt sensitivity in patients with essential hypertension. Jpn Circ J 1987;51:1184–1190.

15 Hinssen H, D'Haese J, Small JV, Sobieszek A: Model of filament assembly of myosins from muscle and nonmuscle cells. J Ultrastruct Res 1978;64:282–302.

16 McCarron DA, Ellison DH, Anderson S: Vasodilation mediated by human PTH 1–34 in the spontaneous hypertensive rat. Am J Physiol 1984;246:F96–F100.

17 Lafferty FW: Primary hyperparathyroidism: changing clinical spectrum, prevalence of hypertension and discriminant analysis of laboratory tests. Arch Intern Med 1981;141:1761–1766.

18 Oshima J, Watanabe M, Hirosumi J, Orimo H: 1,25(OH)$_2$D$_3$ increases cytosolic Ca^{++} concentration of osteoblastic cells, clone MC3T-E1. Biochem Biophys Res Commun 1987;145:956–960.

19   Inoue T, Kawashima H: 1,25-Dihydroxyvitamin D_3 stimulates $^{45}C^{2+}$-uptake by cultured
      vascular smooth muscle cells derived from rat aorta. Biochem Biophys Res Commun
      1988;152:1388–1394.
20   Iseri LT, French JH: Magnesium: natures physiologic calcium blocker. Am Heart J
      1984;108:188–193.
21   Kjeldsen K, Norgaard A: Effect of magnesium depletion on ^{3}H-ouabain binding site
      concentration in rat skeletal muscle. Magnesium 1987;6:55–60.
22   Karaki H, Hatano K, Weiss GB: Effects of magnesium on ^{45}Ca uptake and release at
      different sites in rabbit aortic smooth muscle. Pflügers Arch 1983;398:27–32.
23   Harsselbach W, Fassold E, Migala A, Rauch B: Magnesium dependence of sarcoplasmic
      reticulum calcium transport. Fed Proc 1981;40:2657–2661.

H. Matsuura, MD, 1st Department of Internal Medicine, Hiroshima University School
of Medicine, 1-2-3, Kasumi, Minami-ku, Hiroshima 734 (Japan)

Morii H (ed): Calcium-Regulating Hormones. I. Role in Disease and Aging.
Contrib Nephrol. Basel, Karger, 1991, vol 90, pp 19–24

Abnormality in Sarcoplasmic Reticulum-Dependent Arterial Contraction in Responses to Caffeine and Noradrenaline in Spontaneously Hypertensive Rats

Kyuzo Aoki[a], *Yasuaki Dohi*[a], *Masayoshi Kojima*[a], *Seigo Fujimoto*[b]

[a]Second Department of Internal Medicine and [b]Department of Pharmacology,
Nagoya City University Medical School, Nagoya, Japan

Increased total peripheral vascular resistance in the resting condition has been reported in patients with essential hypertension and rats with gene hypertension [1–3], suggesting that those types of hypertension are caused by the increased vascular resistance. Vascular resistance is mainly determined by the contractile responses of arterial smooth muscle, which are regulated by the cytosolic calcium ion (Ca^{2+}) concentration [4]. The Ca^{2+} concentration is reflected in the ability of the plasma membrane to influx Ca^{2+} and sarcoplasmic reticulum to take up, bind and release Ca^{2+} [5]. It is reported that the sarcoplasmic reticulum may be the only Ca^{2+} storage site in the membrane [5].

Caffeine, one of the vasoconstrictive agents, increases the sensitivity of Ca^{2+}-induced Ca^{2+} release and induces contraction of arterial smooth muscle [6, 7]. Noradrenaline releases Ca^{2+} from the sarcoplasmic reticulum and contracts the muscle [6, 7]. The ability of the sarcoplasmic reticulum could be measured by caffeine- and noradrenaline-induced contractions of the muscles. In the present study, an alteration of Ca^{2+} handling properties of the sarcoplasmic reticulum was investigated in spontaneously hypertensive rat (SHR) [1] arteries by the use of caffeine and noradrenaline. In addition, Ca^{2+} concentration-tension relationship of the contractile proteins was studied in skinned fibres from normotensive Wistar Kyoto rat (WKY) and SHR arteries.

Methods

Tail arteries were isolated from 4-week-old male SHR [1] and age- and sex-matched WKY. The circular arterial strips were mounted horizontally in small muscle chambers to record the isometric tension. Contractile responses to caffeine (10 mmol/l) and noradrenaline $(10^{-5}$ mol/l) were obtained in physiological Ca^{2+} (2.5 mmol/l), low-Ca^{2+} (0.5 mmol/l) and Ca^{2+}-free EGTA (0.1 mmol/l) solution [8]. The obtained contractions were expressed as a percentage of the response to KCl (60 mmol/l). In the other experiment, muscle fibres were skinned with saponin in the mounted strips [8]. The Ca^{2+} concentration-tension relationship was measured in the muscle fibres [8]. Results are presented as mean ± SEM. Data were analysed by Student's t test and $p < 0.05$ was considered significant.

Results

Caffeine Contraction in Low-Ca^{2+} Solution

The caffeine-induced contraction of strip in the physiological Ca^{2+} solution, was significantly smaller in SHR than in WKY. Caffeine did not induce contraction in the Ca^{2+}-free solution. The strips were pre-treated with the Ca^{2+}-free EGTA solution, they then contracted transiently in response to caffeine in the low-Ca^{2+} solution. The contraction was significantly smaller in SHR than in WKY and increased in relation to exposure time in the low-Ca^{2+} solution. The Ca^{2+}-loading duration for the contraction to reach 50% of that (120 s) was significantly greater in SHR (46 ± 3 s) than in WKY (30 ± 3 s, $p < 0.05$) (fig. 1, upper column) [8].

Caffeine Contraction in Ca^{2+}-Free Solution

Caffeine-induced contractions were observed after a few minutes in the Ca^{2+}-free solution, which might be induced by the sarcoplasmic reticulum released Ca^{2+}. The contraction was smaller in SHR than in WKY, indicating that the amounts of caffeine-releasable Ca^{2+} were smaller in SHR than in WKY. The Ca^{2+} unloading time to reach 50% of the contraction (0 min) was significantly shorter in SHR (97 ± 19) than in WKY rats (351 ± 92 s, $p < 0.05$) (fig. 1, middle column) [8].

Fig. 1. a Caffeine (10 mmol/l)-induced contractions (●) were obtained in a physiological Ca^{2+} (2.5 mmol/l) solution (■) and after exposure in a Ca^{2+}-free EGTA 0.1 mmol/l solution (□) caffeine contractions were induced in the low-Ca^{2+} (0.5 mmol/l) solution (▨). The low-Ca^{2+} solution loading time(s) are shown. *b* Caffeine contractions (●) were evoked in the physiological Ca^{2+} solution (■) and after exposure in physiological solution caffeine contractions were induced in the Ca^{2+}-free EGTA solution (□). Ca^{2+} unloading times (min) are shown. *c* Noradrenaline $(10^{-5}$ mol/l)-induced contractions (▲) were obtained in Ca^{2+}-free EGTA solution (□) after incubation with the physiological solution (■). Ca^{2+} unloading times (min) are shown. NA = Noradrenaline. From Dohi et al. [8].

(a)   Caffeine-induced contraction

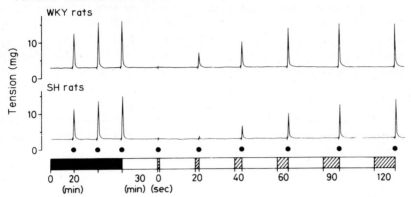

(b)   Caffeine-induced contraction

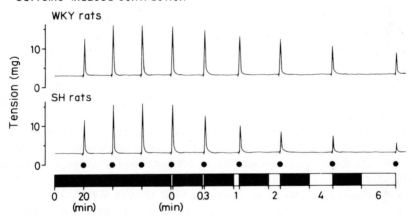

(c)   NA-induced contraction

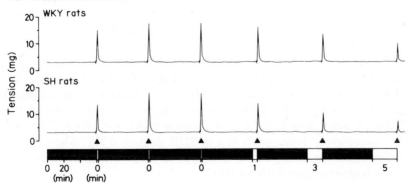

Noradrenaline Contraction in Ca^{2+}-Free Solution

The noradrenaline-induced contraction of the strips after 0 min in the Ca^{2+}-free solution was significantly smaller in SHR than in WKY. The noradrenaline contraction was smaller in SHR than in WKY. The time to reach 50% of the contraction was significantly shorter in SHR (145 ± 8 s) than in WKY (251 ± 26 s, $p < 0.05$) (fig, 1, lower column) [8].

Ca^{2+}-Induced Contraction of Skinned Fibres

The Ca^{2+} concentration-tension relationship in skinned fibres was not different between WKY and SHR. The half-maximal tension was not significantly different between strains [8].

Discussion

Caffeine releases Ca^{2+} from the sarcoplasmic reticulum into the cytosol, increasing cytosolic Ca^{2+} and causing contraction of arterial smooth muscle. Our results have demonstrated that caffeine contraction was smaller in SHR than in WKY strips in a low-Ca^{2+} solution, and that the Ca^{2+} loading time to reach 50% of maximal contraction was longer in SHR than in WKY, suggesting that less Ca^{2+} was taken up by the sarcoplasmic reticulum of SHR than of WKY. However, with prolonged exposure to the low Ca^{2+} solution, the magnitude of contraction was the same for WKY and SHR. Thus, the SHR sarcoplasmic reticulum takes longer than that of WKY to take up the same amount of Ca^{2+}. Moreover, the contraction was smaller in SHR in the Ca^{2+}-free solution and the Ca^{2+} unloading time in the solution to achieve 50% of maximal contraction after 0 min was shorter in SHR than in WKY. This indicates that the amount of caffeine-releasable Ca^{2+} in the sarcoplasmic reticulum is less in SHR than in WKY after limited exposure to the Ca^{2+}-free solution. These data demonstrated that the ability of the sarcoplasmic reticulum to take up and store Ca^{2+} may be diminished in SHR and that the sarcoplasmic reticulum in SHR may release Ca^{2+} rapidly into the cytosol [8–10].

We observed that the noradrenaline-induced contraction was decreased after a prolonged Ca^{2+} unloading, that the contraction in the Ca^{2+}-free solution was smaller in SHR, and that the Ca^{2+} unloading time to reach 50% of the control contraction was shorter in SHR than in WKY. These findings show that the amount of noradrenaline-releasable Ca^{2+} in the sarcoplasmic reticulum is less in SHR than in WKY. The results may indicate that the ability of the sarcoplasmic reticulum to take up and store Ca^{2+} is decreased in SHR [8–10].

Our result that the Ca^{2+} concentration-tension relationship in the skinned fibres from SHR arteries was equal to that from WKY suggests that the Ca^{2+} sensitivity of contractile proteins is the same in WKY and SHR. Therefore, the cause of the differences in caffeine- and noradrenaline-induced contractions between the two strains does not lie in the contractile proteins of the muscle [8].

In conclusion, we have demonstrated that: (1) caffeine-induced contractions of the tail artery in a Ca^{2+}-free solution (containing EGTA) and in a low-Ca^{2+} solution were smaller in SHR than in WKY; (2) noradrenaline-induced contractions of the arteries in the Ca^{2+}-free solution were smaller in SHR than in WKY, and (3) the Ca^{2+} sensitivity of the skinned muscle fibre was the same in both strains. These results suggest that the ability of the sarcoplasmic reticulum to take up and store Ca^{2+} is reduced but that the Ca^{2+} sensitivity of the contractile proteins is normal in SHR arterial smooth muscle. The increased vascular resistance during the development of hypertension in SHR may be caused by the enhanced contraction of the arterial smooth muscle through the reduction to take up and store Ca^{2+} in the sarcoplasmic reticulum [8–10].

Summary

We studied caffeine- and noradrenaline-induced contraction in tail arteries from 4-week-old male SHR and age- and sex-matched WKY. After the sarcoplasmic reticulum Ca^{2+} had been depleted by the Ca^{2+}-free EGTA (0.1 mmol/l) solution, the caffeine (10 mmol/l)-induced contractions in a low-Ca^{2+} (0.5 mmol/l) solution were smaller in SHR than in WKY. After the sarcoplasmic reticulum had been loaded with Ca^{2+} in physiological Ca^{2+} (2.5 mmol/l) solution, caffeine- and noradrenaline (10^{-5} mol/l)-induced contractions in a Ca^{2+}-free EGTA solution were smaller in SHR than in WKY. The Ca^{2+} concentration-tension relationship in skinned arterial fibres was similar in WKY and SHR. These data suggest that the ability of the sarcoplasmic reticulum to take up Ca^{2+} and store Ca^{2+} is decreased in SHR. The decreased take up and store of Ca^{2+} may increase cytosolic Ca^{2+}, which elevates arterial resistance and develops hypertension in gene hypertension.

References

1 Aoki K: The spontaneously hypertensive rat: Evidence of the genetic hypothesis in essential hypertension; in Aoki K, Frohlich ED (eds): Calcium in Essential Hypertension. Tokyo, Academic Press, 1989, pp 3–8.

2 Aoki K, Sato K: Decrease in blood pressure and increase in total peripheral vascular resistance in supine resting subjects with normotension or essential hypertension. Jpn Heart J 1986;27:467–474.

3 Yamamoto J, Nakai M, Natsume T: Cardiovascular responses to acute stress in young-to-old spontaneously hypertensive rats. Hypertension 1987;9:362–370.

4 Bruner CA, Webb RC, Bohr DF: Vascular reactivity and membrane stabilizing effect of calcium in spontaneously hypertensive rats; in Aoki K, Frohlich ED (eds): Calcium in Essential Hypertension. Tokyo, Academic Press, 1989, pp 275–306.
5 Bond M, Somlyo AP: Calcium regulation of contraction of arterial smooth muscle; in Aoki K, Frohlich ED (eds): Calcium in Essential Hypertension. Tokyo, Academic Press, 1989, pp 39–64.
6 Endo M: Calcium release from the sarcoplasmic reticulum. Physiol Rev 1977;55:71–108.
7 van Breemen C, Saida K: Cellular mechanisms regulating $[Ca^{2+}]$i: Smooth muscle. Ann Rev Physiol 1989;51:315–329.
8 Dohi Y, Aoki K, Fujimoto S, Kojima M, Matsuda T: Alteration in sarcoplasmic reticulum-dependent contraction of tail arteries in response to caffeine and noradrenaline in spontaneously hypertensive rats. J Hypertens 1990;8:261–267.
9 Aoki K: The history of the calcium membrane theory of gene (essential) hypertension; in (eds) Essential Hypertension 2. Tokyo, Springer, 1989, pp 373–401.
10 Aoki K, Ikeda N, Yamashita K, Tazumi K, Sato I, Hotta K: Cardiovascular contraction in spontaneously hypertensive rat: Ca^{2+} interaction of myofibrils and subcellular membrane of heart and arterial smooth muscle. Jpn Circ J 1974;38:1115–1121.

Dr. Seigo Fujimoto, Department of Pharmacology, Nagoya City University Medical School, Kawasumi, Mizuho-ku, Nagoya 467 (Japan)

Morii H (ed): Calcium-Regulating Hormones. I. Role in Disease and Aging.
Contrib Nephrol. Basel, Karger, 1991, vol 90, pp 25–35

Calcium Supplementation in Salt-Dependent Hypertension

Komei Saito[a], *Hiroshi Sano*[b], *Yutaka Furuta*[b], *Junji Yamanishi*[b],
Takehiro Omatsu[b], *Yoshihisa Ito*[a], *Hisashi Fukuzaki*[b]

[a]Department of Internal Medicine, Sanda Municipal Hospital, Sanda, Japan;
[b]First Department of Internal Medicine, Kobe University School of Medicine,
Kobe, Japan

It has been well established that excess salt intake increases blood pressure in a subset of patients with essential hypertension [1, 2] and in some models of experimental hypertension [3, 4]. Abnormal renal handling of sodium with the resultant sodium retention due, in part, to an enhanced sympathetic nerve activity may be involved in its mechanism. Recently, clinical [5, 6] and experimental studies [4, 7] have suggested that abnormalities of calcium homeostasis, especially enhanced calcium leak with the resultant calcium deficiency plays an important role in the development of hypertension. In contrast to high sodium intake, intervention studies [8–10] have demonstrated a blood pressure lowering effect of oral calcium supplementation. However, little is known of the effect of oral calcium on the development of salt-dependent hypertension, and the interrelations between sodium and calcium intake in the genesis of hypertension remain unclear. Therefore, the present study investigates the effects of high calcium intake on salt-induced blood pressure elevation and on sodium metabolism in salt-loaded patients with borderline hypertension and in deoxycorticosterone acetate (DOCA)-salt-treated rats.

Materials and Methods

Clinical Study

Twenty-seven untreated patients with borderline hypertension, ages 39–67 years were studied. The patients were admitted and studied for 1 week with their usual diet containing 150 mEq/day sodium, then for 1 week with a low salt diet (50 mEq/day), and finally for 1 week with high salt diet (300 mEq/day). The level of calcium intake remained constant (250 mg/day), and was purposely low to make a definite comparison with the calcium supplementation. During the high sodium period, the subjects were randomly assigned to two treatment groups; the calcium-supplemented patients (Ca group; n = 14) received 6 tablets/

day of 1.0 g calcium glubionate (2,160 mg/day of calcium), and the placebo-treated patients (non-Ca group, n = 13) received the same amount of placebo tablets. Blood pressure and pulse rate were recorded on the 7th day morning of each period by autonomic sphygmo-manometer and blood samples were taken while the patients were supine for 30 min. Ionized calcium levels were measured by the specific calcium ion electrode, and intraerythrocyte contents of sodium (R-Na), potassium (R-K) and magnesium (R-Mg) were determined by our methods described previously [11, 12] with the use of a flame photometry and a polarized atomic absorption spectrometer, respectively. Throughout the study periods, body weight and 24-hour urinary excretion for sodium and potassium were determined every morning.

Experimental Study

Eighty-six male, 6-week-old Wistar-Kyoto rats fed standard rat chow containing 1% calcium were unilaterally nephrectomized with anesthesia. After a week, rats were randomly assigned into one of the following groups; (1) control group: standard chow (1% calcium) and deionized water; (2) DS group: standard chow and 1% NaCl solution for drinking water and treated with DOCA (50 mg/kg s.c.); (3) DS-Ca group: Ca-supplemented chow with CaCl$_2$ (4% Ca) and 1% NaCl solution and treated with DOCA. Systolic blood pressure, heart rate and body weight were measured, and 24-hour urine specimens were collected once a week for 4 weeks. At the end of the experimental period, 46 rats (15 control, 12 DS, and 19 DS-Ca group rats) were killed by decapitation and blood was collected for the measurements of R-Na and ionized Ca level. Hearts were removed after death, quickly frozen and homogenized. Catecholamine concentration of the supernatant was measured by high-performance liquid chromatography. In the other 31 rats, extracellular fluid volume was determined by using sodium thiocyanic acid, and total body sodium concentration by flame photometry after whole-body resolution by nitric acid.

Results

Clinical Study

Figure 1 shows the % change in mean blood pressure level during the course of the low and high salt diets. On the last day of the high salt diet, the % increase in mean blood pressure was significant for the non-Ca group ($+9.69 \pm 2.77\%$, $p < 0.005$) but remained unchanged for the Ca group ($+2.46 \pm 3.17\%$, NS). When compared with the last day of the low salt diet, the % increase in mean blood pressure was smaller in the Ca group than in the non-Ca group ($p < 0.01$). Pulse rate did not show any change. The increase in body weight during high sodium regimen for the Ca group was smaller than that for the non-Ca group (table 1). Serum electrolytes showed small changes after high salt diet, but the degree of the changes was not different between the two groups. As shown in figure 2, the 24-hour urinary sodium excretion was greater in the Ca group on the first and the fourth to the last day of the high salt regimen compared with that in the non-Ca group. The 24-hour sodium/potassium excretion ratio was also higher in the Ca group throughout the high sodium period. R-Na was decreased with salt restriction, and was increased with salt loading in

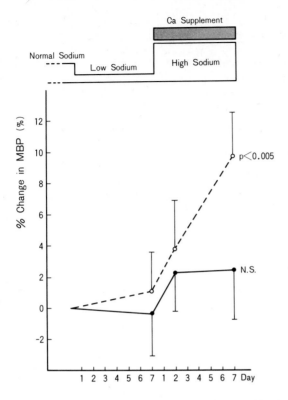

Fig. 1. Line graph showing percent change in mean blood pressure (MBP) during low and high sodium periods for 14 calcium-supplemented (Ca group) patients (●) and 13 placebo (non-Ca group) patients (○). N.S. = Not significant.

both groups (table 1). The increase in R-Na/K during the high sodium regimen was statistically significant in the non-Ca group, but not in the Ca group. The Ca-supplemented patients showed a significant increase in R-Mg on the high salt diet, whereas the placebo-treated ones showed no significant increase. The change in mean blood pressure for the Ca group correlated positively with the change in R-Na/K ($r = 0.62$, $p < 0.02$) and negatively with the increase in R-Mg ($r = -0.54$, $p < 0.05$).

Experimental Study

Systolic blood pressure rose significantly in the DS group from the first to the last week of DOCA-salt treatment, but the development of hypertension was attenuated in the DS-Ca group, as shown in figure 3. Until the last week of the study period, 24 of 32 DS rats died due to unknown reasons,

Table 1. Clinical and laboratory findings, and intraerythrocyte cation contents during three experimental periods

	Calcium-supplemented group			Placebo group		
	normal sodium	low sodium	high sodium	normal sodium	low sodium	high sodium
Age, years		49.2 ± 2.1			52.5 ± 1.5	
Pulse rate, beats/min	58.6 ± 1.5	57.9 ± 1.3	57.1 ± 1.3	60.2 ± 1.9	59.8 ± 2.6	59.8 ± 3.5
Body weight, kg	58.8 ± 1.6	58.2 ± 1.5[c]	58.7 ± 1.5	63.3 ± 3.0	62.5 ± 2.8[c]	68.4 ± 2.8[b]
Difference, %		−0.92 ± 0.39	+0.80 ± 0.20[d]		−1.21 ± 0.35	+1.50 ± 0.30
Serum						
Sodium, mEq/l	139.5 ± 0.5	138.0 ± 0.4	139.4 ± 0.4[a]	139.3 ± 0.5	138.8 ± 0.4	139.3 ± 0.4[a]
Potassium, mEq/l	4.3 ± 0.1	4.3 ± 0.1	4.0 ± 0.1[a]	4.1 ± 0.1	4.1 ± 0.1	3.8 ± 0.1[b]
Calcium, mg/dl	8.5 ± 0.1	8.6 ± 0.1	8.5 ± 0.1	8.6 ± 0.1	8.8 ± 0.2	8.6 ± 0.1[a]
Phosphate, mg/dl	3.3 ± 0.1	3.3 ± 0.1	3.3 ± 0.2	3.2 ± 0.1	3.3 ± 0.1	3.2 ± 0.1
Magnesium, mg/dl	2.1 ± 0.1	2.2 ± 0.1	2.1 ± 0.1	2.1 ± 0.1	2.2 ± 0.1	2.0 ± 0.1[a]
Ionized calcium, mEq/l	2.19 ± 0.05	2.07 ± 0.10	2.10 ± 0.02	2.16 ± 0.03	2.15 ± 0.03	2.14 ± 0.02
RBC						
Sodium, mEq/l · cells	10.36 ± 0.27	9.95 ± 0.27[c]	10.41 ± 0.36[a]	10.43 ± 0.38	9.91 ± 0.36[c]	10.31 ± 0.36[b]
Potassium, mEq/l · cells	100.3 ± 1.5	98.2 ± 1.6	99.0 ± 1.0	98.5 ± 1.2	98.7 ± 1.5	98.1 ± 1.5
Sodium/potassium, ×10⁻¹	1.051 ± 0.034	1.018 ± 0.038[c]	1.053 ± 0.038	1.058 ± 0.036	1.005 ± 0.035[c]	1.052 ± 0.036[a]
Magnesium, mg/dl · cells	6.20 ± 0.40	5.28 ± 0.33[c]	6.65 ± 0.47[b]	6.24 ± 0.75	5.62 ± 0.63	5.89 ± 0.73
Difference, mg/dl · cells		−0.92 ± 0.37	+1.37 ± 0.40[d]		−0.62 ± 0.71	+0.22 ± 0.33

[a] $p < 0.05$, [b] $p < 0.005$ vs. values in low sodium regimen; [c] $p < 0.05$ vs. values in normal sodium regimen; [d] $p < 0.05$ vs. values in placebo group.
Values are mean ± SEM.

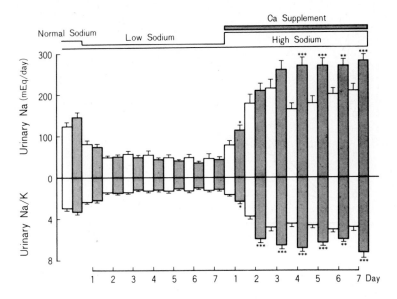

Fig. 2. Bar graph showing daily urinary sodium (Na) excretion (upper bars) and urinary sodium/potassium (Na/K) excretion ratio (lower bars) during the three experimental periods for 14 Ca group patients (shaded bars) and 13 non-Ca group patients (open bars). *p < 0.05, **p < 0.01, ***p < 0.001 vs. values in non-Ca group during the high sodium regimen.

whereas only 1 of 31 calcium-supplemented rats died. Survival rate for the DS-Ca group was significantly higher than that for the DS group (p < 0.02). The 24-hour urinary sodium excretion and urine volume were greater in the DS-Ca rats than in the DS rats from week 1 onwards (fig. 4). Effects of calcium supplementation in the DS rats on sodium metabolism and on catecholamine contents of hearts are recorded in table 2. The mean body weight in the DS-Ca group was not different from that in the DS group. The serum-ionized calcium level for the DS group was significantly lower than that for the control group, but that for the DS-Ca group was not. In the hypertensive DS rats, red cell sodium content, extracellular fluid volume and total body sodium levels were higher (p < 0.01) than those in normotensive control rats. These intra- and extracellular sodium accumulations observed in the DS group were significantly attenuated (p < 0.05–p < 0.005) by calcium supplementation. The concentration of norepinephrine and epinephrine of the hearts was significantly reduced in the DS group (p < 0.01 and p < 0.05, respectively). However, the reduction in catecholamine contents was not observed in the DS-Ca group. Cardiac rate was decreased in the DS-Ca group (p < 0.005) but not in the DS group.

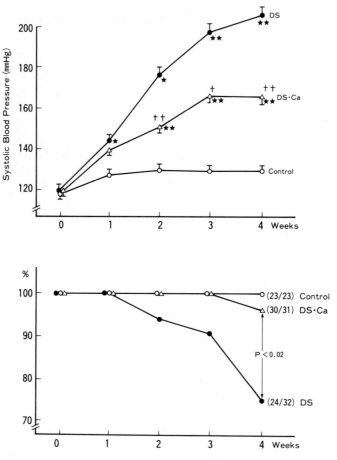

Fig. 3. Line graph showing systolic blood pressure (upper panel) and survival rate (lower panel) of DOCA-salt-treated rats on high (DS-Ca) and regular (DS) calcium diets and of Wistar-Kyoto rats on a regular calcium diet (control). A significant attenuation of blood pressure increase (†p < 0.01, ††p < 0.001) was observed in DOCA-salt rats fed the high (4%) calcium diet (DS-Ca group) when compared to the rats fed a regular calcium diet (DS group); *p < 0.01, **p < 0.001 vs. values in control group. Survival rate for the DS-Ca group was higher than that for the DS group (p < 0.02).

Discussion

Clinical Study

Recent studies in patients with essential hypertension have demonstrated a link between dietary salt intake and calcium, and that the hypotensive efficacy of oral calcium is enhanced in the salt-loaded and

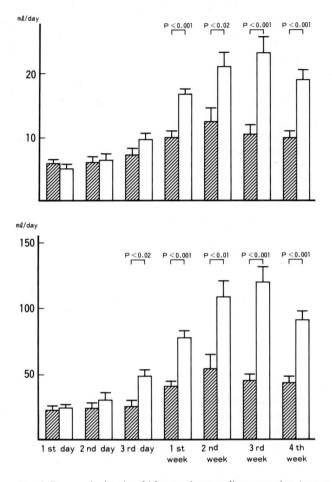

Fig. 4. Bar graph showing 24-hour urinary sodium excretion (upper panel) and urine volume (lower panel) during the experimental period for the DOCA-salt-treated rats on high (4%) calcium diet (open bars) and on regular calcium diet (shaded bars).

low-renin state [13–15]. As we [2] and others [1] have previously reported, approximately half of the salt-loaded patients without calcium supplementation showed a definite salt sensitivity. However, the present results indicate that only 7 days of calcium supplementation while the patient is on a high salt diet can attenuate blood pressure increases, and only 2 of 14 calcium-supplemented patients showed a 10% increase in mean blood pressure.

Table 2. Effects of calcium supplementation in DOCA-salt hypertensive rats

	Controls	DS	DS-Ca
Body weight, g	293 ± 5	225 ± 5[b]	226 ± 3
Serum sodium, mEq/l	139 ± 3	140 ± 3	138 ± 3
Serum calcium, mmol/l	2.82 ± 0.23	2.68 ± 0.21	2.83 ± 0.21
Ionized calcium, mmol/l	1.40 ± 0.02	1.36 ± 0.02[b]	1.40 ± 0.01[d]
Red cell sodium, mEq/l · cells	7.34 ± 0.17	10.92 ± 0.63[b]	9.14 ± 0.37[d]
Extracellular fluid volume, ml/100 g · weight	29.2 ± 0.2	32.9 ± 0.6[b]	31.0 ± 0.5[c]
Total body sodium, mmol/100 g · weight	4.34 ± 0.23	5.74 ± 0.43[b]	5.16 ± 0.13[d]
Cardiac rate, beats/min	459 ± 9	438 ± 10	372 ± 10[d]
Heart-norepinephrine, ng/g	176 ± 14	83 ± 15[b]	189 ± 22[d]
Heart-epinephrine, ng/g	8 ± 2	5 ± 1[a]	7 ± 1[c]

[a] $p < 0.05$, [b] $p < 0.01$ vs. values in control group; [c] $p < 0.05$, [d] $p < 0.005$ vs. values in DS group.
Values are mean ± SEM.

The present balanced study suggests that the greater increase in blood pressure with sodium loading observed in our placebo-treated patients may be attributed to greater sodium retention with a resultant increase in body weight. It is possible that an extremely low calcium intake (250 mg/day) and a calcium-deficient state may be responsible for the impaired natriuresis in the non-Ca group. As observed in the calcium-supplemented group, urinary sodium excretion may increase with increasing calcium excretion [13, 15], since calcium and sodium excretion are related to each other for common transport pathways for reabsorption in the proximal tube and the loop of Henle [16]. The prevention of the marked sodium retention, and hence a smaller weight gain may, at least in part, contribute to the attenuation of blood pressure rise with salt loading.

Recent reports suggest that suppression in membrane active transport for sodium with a resultant increase in intracellular sodium may be involved in the mechanism of the volume-expanded form of human hypertension [17]. In the present study, since the degree of accumulation of cellular sodium during the high salt regimen was not different between the two groups, it may be difficult to conclude that the hypotensive action of oral calcium was mediated to the attenuation of salt-induced intracellular sodium retention. An increase in intraerythrocyte magnesium content was observed in the Ca group and was significantly greater than that in the non-Ca group. Although the mechanism remains unclear at the moment, this is consistent with the finding of Resnick et al. [10] that a high calcium diet induces an elevation in the intraerythrocyte free magnesium level by more than 35%. We have previously reported that the hypotensive action of oral magnesium was associated with an increase in red cell magnesium [11]. The relation between the increase in cellular magnesium and the attenuation of blood pressure rise in our calcium-supplemented patients suggest that the change in intracellular magnesium may be involved in the regulation of blood pressure response to salt loading.

Experimental Study

The present experimental study clearly indicates that dietary calcium supplementation given as 4% $CaCl_2$ can attenuate the development of hypertension in DOCA-salt-treated rats. The reduced blood ionized calcium level in our DS group is consistent with the findings of Wright et al. [18], suggesting the altered calcium metabolism in the development of mineral corticoid-salt dependent hypertension. Although the mechanism cannot be elucidated, the observed effects of oral calcium on blood pressure and on calcium metabolism parallel previous observations in spontaneously hypertensive rats [7] or Wistar-Kyoto rats [19].

The increase in extracellular fluid volume, total body sodium accumulation and the rise in red cell sodium content in our DS rats may represent extra- and intracellular sodium retention which contribute to the development of a salt-dependent, volume-expanded type of hypertension. The high calcium diet in this study resulted in an enhanced natriuresis throughout the experimental period that could account for the normalization of sodium retention in DS rats. The calcium-induced natriuresis is consistent with the findings of our clinical study [20], and, theoretically, the well-known close association between the handling of sodium and calcium at the renal proximal tubule may be responsible for the enhanced natriuresis in the DS-Ca group as suggested by Barry and Lehman [21].

An enhanced activity of the peripheral sympathetic nervous system and a decrease in the content of norepinephrine of the heart have been reported in DOCA-salt hypertensive rats [3]. The faster disappearance of endogenous norepinephrine could be attributed to an increased release due to enhancement of nerve impulse flow. Calcium-induced natriuresis might attenuate the decrease of cardiac catecholamine contents in DOCA-salt rats by suppressing enhanced sympathetic nerve activity due, in part, to sodium retention. The decreased heart rate in the DS-Ca group may be attributable to this inhibited sympathetic nerve activity.

Summary

To clarify the mechanism of the antihypertensive effect of oral Ca loading, we studied the effect of Ca supplementation on salt-induced blood pressure elevations in patients with essential hypertension and DOCA-salt hypertensive rats. When the diet was changed from low to high salt (300 mEq/day), the percent increase in mean blood pressure was smaller (p < 0.01) in the Ca-supplemented (2,160 mg/day) patients than in the Ca-restricted (250 mg/day) ones. Oral Ca loading resulted in a smaller weight gain, a greater urinary sodium excretion, and an increase in red cell Mg. In the experimental study, high Ca (4% $CaCl_2$) intake attenuated the blood pressure elevation in DOCA-salt-treated rats, accompanied with an increase in urinary sodium excretion, with the resultant attenuation in intra- and extracellular sodium retention. The decrease in catecholamine contents of hearts was improved, and a higher survival rate was observed in Ca-supplemented DOCA-salt rats. The results suggest that Ca supplementation may prevent a rise in BP in salt-dependent hypertension by inducing natriuresis with the resultant attenuation in sodium retention. The altered intracellular Mg level in hypertensive patients and the normalization of enhanced sympathetic nervous activity in DOCA-salt rats may, in part, be involved in its mechanism.

References

1   Fujita T, Henry WL, Bartter FC, Lake CR, Delea CS: Factors influencing blood pressure in salt-sensitive patients with hypertension. Am J Med 1980;69:334–344.
2   Saito K, Furuta Y, Sano H, Okishio T, Fukuzaki H: Abnormal relationship between dietary sodium intake and red cell sodium transport in salt-sensitive patients with essential hypertension. Clin Exp Hypertens 1985;7:1217–1232.
3   De Champlian J, Krakoff LR, Axelrod J: Catecholamine metabolism in experimental hypertension in the rat. Circ Res 1967;20:136–145.
4   Umemura S, Smyth DD, Nicar M, Rapp JP, Pettinger WA: Altered calcium homeostasis in Dahl hypertensive rats: Physiological and biochemical studies. J Hypertens 1986;4:19–26.
5   McCarron DA, Morris CD, Henry HJ, Stanton JL: Blood pressure and nutrient intake in the United States. Science 1984;224:1392–1398.
6   Strazzullo P, Nunziata V, Cirillo M, Giannattasio R, Mancini M: Abnormalities of calcium metabolism in essential hypertension. Clin Sci 1983;65:137–141.
7   McCarron DA, Yung NN, Ugoretz BA, Krutzik S: Disturbance of calcium metabolism in the spontaneously hypertensive rat. Hypertension 1981;3(suppl 1):162–167.
8   McCarron DA, Morris CD: Blood pressure response to oral calcium in persons with mild to moderate hypertension. A randomized, double blind, placebo-controlled, crossover trial. Ann Intern Med 1985;103:825–831.
9   Kurtz TW, Morris RC: Attenuation of deoxycorticosterone-induced hypertension by supplemental dietary calcium. J Hypertens 1986;4(suppl 5):129–131.
10 Resnick LM, Gupta RK, Sosa RE, Corbett ML, Sealey JE, Laragh JH: Effects of altered dietary calcium intake in experimental hypertension: Role of intracellular free magnesium. J Hypertens 1986;4(suppl 5):182–185.
11 Saito K, Hattori K, Omatsu T, Hirouchi H, Sano H, Fukuzaki H: Effects of oral magnesium on blood pressure and red cell sodium transport in patients receiving long-term thiazide diuretics for hypertension. Am J Hypertens 1988;1:71S–74S.
12 Saito K, Furuta Y, Omatsu T, Ooshima T, Nishimura Y, Takano S, Fukuzaki H, Okishio T, Sano H, Hirouchi H: Relationship between red cell cation contents and blood pressure level. Jpn Heart J 1985;26:955–964.

13   Zemel MB, Gualdoni SM, Sowers JR: Sodium excretion and plasma renin activity in normotensive and hypertensive black adults as affected by dietary calcium and sodium. J Hypertens 1986;4(suppl 6):343–345.

14   Resnick LM: Uniformity and diversity of calcium metabolism in hypertension. A conceptual framework. Am J Med 1987;82(suppl 1B):16–26.

15   Lasaridis AN, Kaisis CN, Zananiri KI, Syrganis CD, Tourkantonis AA: Increased natriuretic ability and hypotensive effect during short-term calcium intake in essential hypertension. Nephron 1989;51:517–523.

16   Antoniou LD, Eisner GM, Slotkoff LM, Lilienfield LS: Relationship between sodium and calcium transport in the kidney. J Lab Clin Med 1969;74:410–420.

17   Haddy FJ: Abnormalities of membrane transport in hypertension. Hypertension 1983;5(suppl 5):66–72.

18   Wright GL, Rankin GO: Concentrations of ionic and total calcium in plasma of four models of hypertension. Am J Physiol 1982;243:H365–H370.

19   McCarron DA: Blood pressure and calcium balance in the Wistar-Kyoto rat. Life Sci 1982;30:683–689.

20   Saito K, Sano H, Furuta Y, Fukuzaki H: Effect of oral calcium on blood pressure response in salt-loaded borderline hypertensive patients. Hypertension 1989;13:219–226.

21   Barry GD, Lehman HH: Effect of increased dietary calcium on the development of experimental hypertension. Fed Proc 1977;36:492.

Komei Saito, MD, Internal Medicine, Sanda Municipal Hospital,
1-11, Yashiki-machi, Sanda-shi, 669-13 (Japan)

Morii H (ed): Calcium-Regulating Hormones. I. Role in Disease and Aging.
Contrib Nephrol. Basel, Karger, 1991, vol 90, pp 36–41

Antihypertensive Effects of Oral Calcium Supplementation in Spontaneously Hypertensive Rats

Takuzo Hano, Akira Baba, Jiro Takeda, Ichiro Nishio,
Yoshaiki Masuyama

Division of Cardiology, Department of Medicine, Wakayama Medical College,
Wakayama, Japan

Recent publications have verified the finding of an inverse relationship between dietary calcium intake and blood pressure [1, 2]. And it has been shown that an increased calcium intake suppressed systolic blood pressure in spontaneously hypertensive rats (SHR; Okamoto and Aoki) [3]. However, the precise mechanisms of antihypertensive action by calcium supplementation were obscure. Since our recent studies showed the enhanced pressor response to norepinephrine (NE), the increased cytosolic free calcium concentration and the reduced membrane fluidity in SHR, the present study was designed to clarify whether increased dietary calcium intake affects these vascular and cellular abnormalities in SHR.

Materials and Methods

Male 4-week-old SHR and age-matched Wistar-Kyoto rats (WKY) were divided into two groups, the control group and the calcium-supplemented group. In the calcium-supplemented group, 1.2% $CaCl_2$ was given as drinking water ad libitum for 3 weeks. Distilled water was given to the control group. Blood pressure was measured every week by the tail-cuff plethysmographic method. After being on the diets for 3 weeks, each rat was anesthetized with sodium pentobarbital (60 mg/g) to take a blood sample and to make a mesenteric arterial preparation.

Measurement of Plasma Norepinephrine Concentration

To measure the plasma NE concentration, a polyethylene catheter was inserted into the right atrium through a jugular vein 1 day prior to sacrifice. 500 μl of blood was taken out from the catheter in a conscious and unrestricted condition. Plasma NE was determined by high-pressure liquid chromatography with an electrochemical detector.

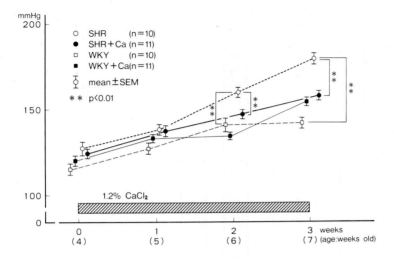

Fig. 1. Effects of calcium supplementation on systolic blood pressure.

Mesenteric Arterial Preparation

A mesenteric artery-intestine loop preparation was made as previously reported and was perfused with Krebs-Henseleit solution at a constant flow of 3 ml/min [4]. Pressure responses to oxogeneous NE were tested by a bolus injection of $0.5-2.0\ \mu g$ of NE to the arterial cannula.

Measurement of Intracellular Cytosolic Free Calcium

Platelets were prepared from platelet-rich plasma using a Sepharose CL-2B column. Platelets were incubated with $1\ \mu M$ Fura-2/AM for 60 min at 25 °C. After centrifugation, platelets were resuspended in 145 mM NaCl, 5 mM KCl, 0.5 mM Na$_2$HPO$_4$, 1 mM MgSO$_4$, 5 mM glucose, and 10 mM Hepes buffer (pH 7.4) containing 1 mM CaCl$_2$ or 1 mM EGTA. Fluorescence was measured at 340 and 380 nm excitation and at 500 nm emission wavelength [5, 6]. Ca^{2+} concentration was calculated using the formula described by Grynwicz et al [7].

Statistical Analysis

All data were expressed as mean $\pm$ SEM. Comparison between two means of independent samples means was made with the unpaired Student's t test. Multiple comparisons were made with Bonferroni's method after analysis of variance. A probability level of p below 0.05 was considered significant.

Results

The elevation of blood pressure was suppressed by high calcium supplementation. A significant difference was obtained 2 weeks after calcium supplementation. In contrast, WKY showed no significant fall in blood pressure (fig. 1).

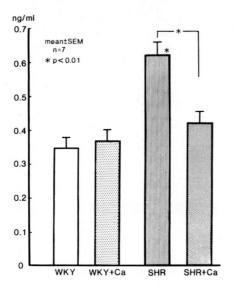

Fig. 2. Effects of calcium supplementation on plasma NE concentration.

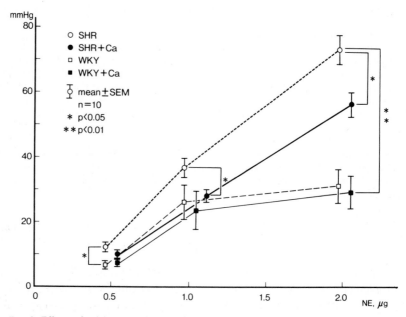

Fig. 3. Effects of calcium supplementation on pressor responses to NE.

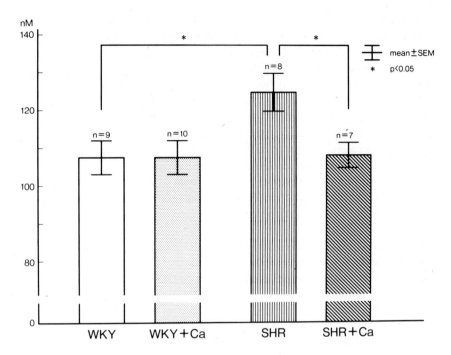

Fig. 4. Effects of calcium supplementation on intracellular free calcium concentration in platelets.

Water intake and urinary sodium excretion were unchanged by oral calcium supplementation in both SHR and WKY.

Plasma NE concentrations were significantly higher in SHR than that in control WKY. Calcium supplementation reduced the plasma NE level in SHR to almost the same levels as the controls, but not in WKY (fig. 2).

Calcium supplementation attenuated the increased pressor response to exogenous NE in isolaged mesenteric artery of SHR (fig. 3).

Intracellular free calcium concentrations were higher in SHR than those in WKY. Ca cupplementation suppressed the intracellular calcium levels in SHR to the control levels (fig. 4).

Discussion

The SHR is characterized by the increased sympathetic nerve activity and vascular responsiveness, especially in the early stage [4]. The calcium antagonist, verapamil, suppressed NE overflow and pressor response in the isolated mesenteric arterial preparation. These reductions of pressor response and NE overflow were more marked in SHR compared with WKY [8]. These results indicate that the calcium influx plays an important role in the augmentation of NE overflow and pressor response in SHR. Since platelet is one of the useful models of vascular smooth muscle cells, we measured the cytosolic free calcium contentration of platelets, which was increased at any age in SHR [5]. Membrane fluidity of erythrocytes was reduced in SHR and was reduced more markedly in SHR by calcium loading with calcium ionophore than the controls [9]. These results suggest that there is altered calcium handling in the cellular levels in SHR, which causes the increased sympathetic and vascular tone.

In the present study, calcium supplementation reduced the pressor response to NE in the isolated mesenteric arterial preparation. The effects of calcium supplementation on platelet intracellular calcium concentrations were evaluated to clarify the mechanisms of the suppressed vascular responsiveness. Intracellular free calcium concentrations in platelets were normalized by calcium supplementation in SHR.

These data suggest that calcium supplementation normalized calcium handling at the cellular levels. As a result of normalized intracellular calcium concentration, vascular responsiveness is suppressed and blood pressure is reduced.

Increased plasma NE levels shown in SHR were also suppressed by calcium supplementation to almost the same levels as WKY. Calcium supplementation might suppress the sympathetic tone.

In conclusion, calcium supplementation normalizes the altered intracellular calcium regulation in SHR, and attenuates the sympathetic and vascular tones, resulting in a blood pressure fall in SHR.

Summary

This study was designed to clarify the mechanisms underlying the antihypertensive action of oral calcium supplementation in SHR. Four-week-old SHR and age-matched WKY were divided into calcium-supplemented and control groups. Calcium supplementation was carried out by giving 1.2% $CaCl_2$ solution as drinking water ad libitum. Distilled water was given to the control group. After 3 weeks of treatment, plasma NE, pressor response to NE in isolated mesenteric artery, platelet cytosolic free calcium concentration and membrane fluidity of erythrocytes were evaluated. The elevation of blood pressure were retarded in calcium supplemented SHR. Calcium supplementation reduced the augmented pressor response to NE and the high level of cytosolic free calcium concentration in SHR. WKY showed no significant changes of these parameters by calcium supplementation. In conclusion, calcium suplementation reduces blood pressure through the reduction of sympathetic and vascular tone in SHR.

References

1 Stitt FW, Clayton DG, Crawford MD, Morris JN: Clinical and biochemical indicators of cardiovascular disease among men living in hard and soft water areas. Lancet 1973;i:122–126.

2 McCarron DA, Morris CD, Cole C: Dietary calcium vs. human hypertension. Science 1982;217:267–269.

3 Ayachi S: Increased dietary calcium lowers blood pressure in the spontaneously hypertensive rat. Metabolism 1979;28:1234–1238.

4 Hano T, Rho J: Norepinephrine overflow in perfused mesenteric arteries of spontaneously hypertensive rats. Hypertension 1989;14:44–53.

5 Baba A, Fukuda K, Kuchii M, Ura M, Yoshikawa H, Hamada M, Hano T, Nishio I, Masuyama Y.: Intracellular free calcium concentration, Ca^{2+} channel and calmodulin level in experimental hypertension in rats. Jpn Circ J 1987;51:1216–1222.

6 Baba A, Fukuda K, Hano T, Shiotani M, Yoshikawa K, Ura M, Nakamura Y, Kuchii M, Nishio I, Masuyama Y: Responses of cytosolic free calcium to ADP in platelets of spontaneously hypertensive rats. Am J Hypertens 1990;3:2065–2095.

7 Grynkiewicz G, Poinie M, Tsien R: A new generation of Ca^{2+} indicators with greatly improved fluorescence properties. J Biol Chem 1985;260:3440–3450.

8 Masuyama Y, Tsuda K, Kuchii M, Nishio I. Effects of diltiazem (a calcium antagonist) on neurosecretion and vascular responsiveness in hypertension. J Hypertens 1986;3(suppl 3):s169–172.

9 Tsuda K, Tsuda S, Minatogawa Y, Iwahashi H, Kido Y, Masuyama Y: Decreased membrane fluidity of erythrocytes and cultured vascular smooth muscle cells in spontaneously hypertensive rats: An electron spin resonance study. Clin Sci 1988;75:477–480.

Dr. Takuzo Hano, Division of Cardiology, Department of Medicine,
Wakayama Medical College, 7-Bancho 27, Wakayama, 640 (Japan)

Morii H (ed): Calcium-Regulating Hormones. I. Role in Disease and Aging.
Contrib Nephrol. Basel, Karger, 1991, vol 90, pp 42–48

Calcium Uptake into Enterocyte Brush-Border Membrane Vesicles is Greater in Spontaneously Hypertensive than in Normotensive Control Rats[1]

U. Hennessen[a], *L. Comte*[a], *M.-C. Steuf*[a], *B. Lacour*[a], *T. Drüeke*[a], *D.A. McCarron*[b]

[a]Inserm Unité 90, Départment de Néphrologie, Hôpital Necker, Paris, France;
[b]Division of Nephrology and Hypertension, Oregon Health Sciences University, Portland, Oreg., USA

Numerous reports indicate that calcium (Ca) plays an important role in human as well as in experimental hypertension [1–3]. Alterations of Ca metabolism have been particularly well described in the spontaneously hypertensive rat (SHR) compared with its normotensive control rat, the Wistar Kyoto rat (WKY) [4–7]. Furthermore, feeding a Ca-enriched diet (2% by weight) to the young SHR prevented the emergence of hypertension [8, 9]. We [7, 10] and others [11] have found a decrease in intestinal Ca absorption in the mature SHR in vivo and in vitro, even though this finding is still matter of debate [3, 6, 12]. A defect in intestinal Ca transport has been observed in different experimental settings [3, 6]. However, even the use of isolated cells [13, 14] did not allow the site of the anomaly to be localized. Therefore, we studied Ca uptake in isolated duodenal brush-border membrane vesicles (BBMV) in SHR and WKY rats.

Materials and Methods

Animals. Male SHR and WKY were obtained at the age of 8 weeks from IFFA CREDO (Centre de recherche et élevage des Oncins, l'Arbesle, France). During 4–6 weeks they were fed a synthetic diet containing Ca 1%, P 0.46%, and vitamin D_2 2,200 U/kg food (Centre National de Recherches Zootechniques, la Minière par Versailles, France). Animals had free access to food and water. They were fasted during the 18 h before sacrifice, which took place at the age of 12–14 weeks.

[1]This work was supported by grants-in-aid from the Comité des Salines, France, and the National Dairy Council, United States.

Preparation of Brush-Border Membrane Vesicles. Under a light ethyl-ether anesthesia the duodenum was excised and rinsed twice with ice-cold saline and gently scraped. Four rats were used for each vesicle-preparation and vesicles from SHR and WKY were prepared the same day in parallel. We used the method described by Forstner et al. [15] with slight modifications. Briefly, mucosa was suspended in chilled buffer I (20 mM Hepes, 5 mM Na$_3$EDTA, 0.1 mM PMSF at pH 7.4) and homogenized by a Polytron. After centrifugation for 10 min at 3,000 g, the supernatant was discarded and the pellet was resuspended in buffer I and recentrifugized for 10 min at 3,000 g. That step was repeated twice and the pellet was suspended in chilled buffer II (20 mM Hepes, 0.8 mM Na$_3$EDTA, 90 mM NaCl, pH 7.4). After a 30-min rest at 4 °C, the suspension was filtered twice. The liquid was spun for 10 min at 3,000 g. The pellet was resuspended in buffer III (100 mM mannitol, 20 mM HEPES, 0.1 mM PMSF, pH 7.4). After 20 min centrifugation at 24,000 g the pellet was collected and homogenized in a glass-Teflon Potter and MgSO$_4$ was added to buffer III to attain a final concentration of 1 mM. The brush-border vesicles were contained in the pellet resulting from a second centrifugation at 24,000 g.

Enzyme Measurements. Sucrase activity was measured according to the technique described by Dahlquist [16] and Na$^+$,K$^+$-ATPase activity was measured as described by Scharschmidt et al. [17] and defined as the ouabain-sensitive hydrolysis of ATP. Protein was measured as described by Lowry et al. [18] using bovine albumin as standard.

Uptake Studies. Buffer III was used as the incubation solution. About 150 μg of BBMV protein were incubated at 25 °C during 10 s in a solution with varying Ca^{2+} concentrations (0.025–1.0 mM) to which radiolabelled ^{45}Ca was added. Reaction was stopped by the addition of chilled stop solution containing (in mM) 100 mannitol, 20 HEPES, 75 MgCl$_2$, and 1 EGTA at pH 7.4 and the samples were immediately filtered under suction (Millipore filters, type HA, 0.45 μm, 25 mm in diameter). The filter was washed with 7.5 ml stop solution. Radioactivity remaining on the filter was determined by liquid scintillation after dissolving filter in 5 ml of Pico Fluor 30 (Packard). Kinetic constants were derived from each individual experiment. Kinetics of saturable and nonsaturable Ca transport were calculated as described by Wilson et al. [19]. V$_{max}$ and K$_m$ were calculated for each preparation by the method of Lineweaver and Burk [20]. Statistical analyses were realized by Student's t test.

Materials. N-2-hydroxyethylpiperazine N'-2 ethanesulfonic acid (HEPES), phenyl-methylsulfonyl-fluoride (PMSF), ethyleneglycol-bis-β-aminoethylether NNN'N'tetraacetic acid (EGTA), ethylenediamine-tetraacetic acid trisodium salt (Na$_3$EDTA) and other chemicals were from Sigma Chemicals (St Louis, Mo).

Results

We used sucrase activity as a marker for the brush-border membrane and Na$^+$,K$^+$-ATPase activity as a marker for the basolateral membrane. Sucrase activity enrichment was 12-fold for both strains and Na$^+$,K$^+$-ATPase activity was 0.6-fold diminished in WKY (table 1). These values correspond to values reported in the literature [21, 22]. In BBMV preparations of SHR we were unable to detect duodenal Na$^+$,K$^+$-ATPase activity even when high concentrations of ouabain were used (10 mM), and we

Table 1. Enzyme marker activity

Homogenate	WKY, U/mg protein	SHR, U/mg protein
Homogenate	0.23 ± 0.05	0.23 ± 0.08
BBMV	2.76 ± 0.97	2.70 ± 0.82

Enzyme marker activities were measured in homogenate and in final duodenal BBMV of SHR and WKY. Values are the results of 6 separate preparations and expressed as means $\pm$ SD.

found just a weak activity in WKY, indicating only minor contamination by basolateral membrane. Total Ca^{2+} uptake by isolated BBMV is shown in figure 1. It tended to be greater in SHR than in WKY but the difference was not statistically significant. Similarly, non-saturable uptake was comparable in both strains (fig. 1). Saturable Ca^{2+} uptake, however, was greater in SHR than in WKY, as illustrated in figure 2. When expressed by Michaelis-Menten analysis V_{max} was significantly higher in SHR than in WKY (table 2). Values for K_m did not differ significantly between the two strains.

Discussion

The novel finding of the present study was an increase in the V_{max} of BBMV-mediated Ca^{2+} transport in the SHR, compared with that of the WKY. Whether this increase is due to a greater number of transporters or a higher turnover, could not be clarified by our experimental setting. The apparent affinity (K_m) of this transport system for Ca^{2+} was not different between the two strains. We used exclusively the duodenum which represents the site of maximal active Ca^{2+} transport [23]. BBMV were essentially pure, closed, and right-side-out orientated, as shown by electron microscopy. Sucrase activity enrichment of isolated BBMV corresponded to values found by other groups [21, 22]. Moreover, determinations of Na^+,K^+-ATPase activity indicated only minor contamination by basolateral membranes (BLM). All experiments of BBMV Ca^{2+} uptake were done at the time point of 10 s after addition of Ca^{2+}. In contrast to other authors who chose time points at 3 s [19] or 7 s [24], our experiments showed linear uptake up to 30 s (data not shown). We could clearly separate saturable from nonsaturable transport of Ca^{2+} into duodenal BBMV. Nonsaturable 'uptake' corresponds to diffusive transport of Ca^{2+} into the vesicular space and nonspecific binding to the outer surface of

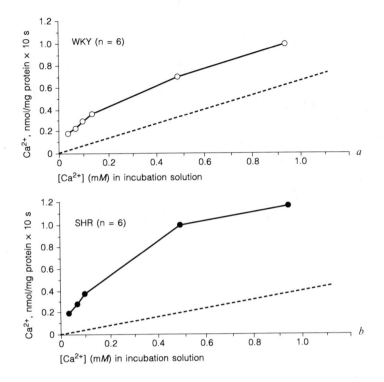

Fig. 1. Total Ca^{2+} uptake by duodenal BBMV of WKY and SHR. Total Ca^{2+} uptake as described in Materials and Methods by duodenal BBMV of WKY (*a*) and SHR (*b*) was measured as with increasing Ca^{2+} concentration (0.036; 0.069; 0.1; 0.13; 0.49; 0.94 m*M*) in the incubation solution. Presented are means of 6 separate experiments in duplicate for each strain of rats. Calculated nonsaturable uptake.

vesicles, as shown by experiments done at 4 °C (data not shown). The existence of a saturable uptake implies a mediated transport system which we shall call transporter. Experiments exploring the interaction of other divalent cations with this tranporter seem to indicate a mobile carrier [19].

The present results of enhanced mediated Ca^{2+} uptake by duodenal BBMVs in the SHR have to be interpreted in the light of our previous findings in isolated enterocytes [13, 14] which indicated a decreased whole cell Ca^{2+} uptake in the 12- to 14-week-old male SHR compared with the WKY of same age and gender. It is conceivable that Ca^{2+} penetrates more rapidly into the cell because of abnormal brush-border membrane (BBM) structure, recently demonstrated by electron microscopy [25]. Alternatively,

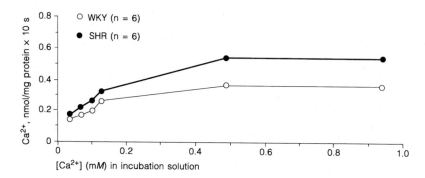

Fig. 2. Saturable Ca^{2+} uptake by duodenal BBMV of WKY and SHR. Saturable uptake of Ca into duodenal BBMV of WKY and SHR was calculated by reducing total Ca uptake by nonsaturable uptake. Presented are means of 6 separate experiments in duplicate for each group of rats.

Table 2. Kinetic constants of the saturable Ca^{2+} uptake in BBMV

	WKY	SHR
V_{max}	0.346 ± 0.097	0.576 ± 0.186**
K_m	0.058 ± 0.049	0.103 ± 0.024

Values are expressed as means $\pm$ SD, V_{max} and K_m were calculated as described in 'Materials and Methods'. ** $p < 0.05$.

a greater number of Ca^{2+} channels could be open at the site of the BBM level in response to an impaired Ca^{2+} extrusion at the site of the BLM, leading to an increase in cytosolic free Ca^{2+} and thereby an activation of the Ca^{2+}-ATPase via the Ca^{2+}-calmodulin complex.

For technical reasons, the determination of enterocyte cytosolic free Ca^{2+} is still difficult. In preliminary experiments, it has been estimated to be approximately 130 nmol/l in the intestinal villus epithelial cell [26]. In the SHR or WKY enterocyte, a precise measurement of cytosolic free Ca^{2+} has not yet been achieved. On the other hand, determinations in erythrocytes and platelets from hypertensive subjects and animals [27, 28] have revealed an increase of free cytosolic Ca^{2+} which could be due to alterations of the kinetic properties of the plasma membrane Ca^{2+}-ATPase pump [29]. This might also be true for enterocytes. Finally, it is also conceivable that Ca^{2+}, immediately after its entry into the SHR's enterocyte, is bound or transported into structures from which it is less accessible than in the WKY for outward transport across the BLM. Taken together, our present and previous findings point to a perturbation of the Ca^{2+} entry and extrusion steps in the intestinal epithelium of the SHR. The limiting step seems to be localized at the basolateral level.

Summary

Arterial hypertension in the SHR is associated with disturbances of calcium homeostasis, compared with its normotensive control, the WKY. In order to study intestinal Ca^{2+} handling at the subcellular level, we examined $^{45}Ca^{2+}$ uptake kinetics in isolated brush-border membrane vesicles (BBMV). Experiments were conducted in male, 12- to 14-week-old rats on a 1% Ca diet. BBMV were purified by the method of Forstner et al. [15]. No difference in BBMV enrichment was observed between SHR and WKY. Ca^{2+} uptake was studied at various Ca^{2+} concentrations in the incubation medium ($0.025-1.0\,mM$) and could be separated into a nonsaturable and a saturable component. The saturable component followed Michaelis-Menten kinetics. V_{max} in the SHR was greater than in the WKY: 0.576 ± 0.186 (n = 6) vs. 0.346 ± 0.10 nmol/mg protein $\times$ 10 s (n = 6), mean $\pm$ SD, $p < 0.05$. However, K_m was not different in the two animal strains. In conclusion, mediated Ca^{2+} transport into duodenal BBMV was increased in the adolescant SHR. When considering that the transcellular duodenal Ca^{2+} flux is decreased in the SHR at this age, the rate-limiting step of perturbed transeptithelial Ca^{2+} transport is probably localized at the site of the basolateral membrane.

References

1 McCarron DA, Morris CD, Cole C: Dietary calcium in human hypertension. Science 1982;217:267–269.

2 Yamamoto ME, Kuller KH: Does dietary calcium influence blood pressure? Evidence from the three stroke mortality study (1971–1974). Circulation 1985;72(suppl 3):116.

3 Bukoski RD, McCarron DA: Calcium and hypertension; in Baker PF (ed): Handbook of Experimental Pharmacology. Berlin, Springer, 1988, vol 83, pp 467–487.

4 McCarron DA: Calcium metabolism and hypertension. Kidney Int 1989;35:717–736.

5 Resnick LM: Calcium metabolism in pathophysiology and treatment of clinical hypertension. Am J Hypertens 1989;2:179S–185S.

6 Drüeke T, Bourgouin P, Lacour B: Disturbances of calcium metabolism in experimental hypertension. Miner Electrolyte Metab 1990;16:6–11.

7 Bourgouin P, Lucas P, Roullet C, Pointillart A, Thomasset M, Brami M, Comte L, Lacour B, Garabédian M, McCarron DA, Drüeke T: Developmental changes of Ca^{2+}, PO_4, and cacitriol metabolism in spontaneously hypertensive rats. Am J Physiol. 1990;259:F104–F110.

8 MaCarron DA, Lucas P, Shneidman J, Lacour B, Drüeke T: Blood pressure development of the spontaneously hypertensive rat following concurrent manipulations of dietary Ca^{2+} and Na^+. Relation to intestinal Ca^{2+} fluxes. J Clin Invest 1986;76:1147–1154.

9 Hatton DC, Scrogin KE, Metz JA, McCarron DA: Dietary calcium alters blood pressure reactivity in spontaneously hypertensive rats. Hypertension 1989;13:622–629.

10 Lucas PA, Brown RC, Drüeke T, Lacour B, Metz G, McCarron AD: Abnormal Vitamin D metabolism, intestinal Ca transport, and bone calcium status in the spontaneously hypertensive rat compared with its genetic control. J Clin Invest 1986;78:221–227.

11   Schedl HP, Wilson HD, Horst RL: Calcium transport and vitamin D in the three breeds of spontaneously hypertensive rats. Hypertension 1988;12:310–316.

12   Bindels RJM, van den Brock LAM, Jongen MJM, Hackeng WHL, Löwick CW, van Os CH: Increased plasma calcitonin levels in young spontaneously hypertensive rats: Role in disturbed phosphate homeostasis. Pflügers Arch 1987;408:395–400.

13   Lucas PA, Roullet CM, Duchambon P, Lacour B, Dang P, McCarron DA, Drüeke T: Decreased duodenal enterocyte calcium fluxes in the spontaneously hypertensive rat. Am J Hypertens 1989;2:86–92.

14   Roullet C, Young EW, Roullet J-B, Lacour B, Drüeke T, McCarron DA: Calcium uptake by duodenal enterocytes isolated from young and mature SHR and WKY rats: influence of dietary calcium. Am J Physiol 1989;257:F574–F579.

15   Forstner GG, Sabesin SM, Isselbacher KJ: Rat intestinal microvillus membranes; purification and biochemical characterization. Biochem J 1968;106:381–390.

16   Dahlquist A: Methods for the assay of intestinal disaccharidases. Analyt Biochem 1964;7:18–25.

17   Scharschmidt BF, Keeffe EB, Blankenship NM, Ockner RK: Validation of a redording spectrophotometric method for measurement of membrane-associated Mg- and NaK-ATPase activity. J Lab Clin Med 1979;93:790–799.

18   Lowry OH, Rosenbrough NJ, Farr AL, Randall RJ: Protein measurement with the Folin phenol reagent. J Biol Chem 1951;193:256–275.

19   Wilson HD, Schedl HP, Christensen K: Calcium uptake by brush-border membrane vesicles from the rat intestine. Am J Physiol 1989;26:F446–F453.

20   Lineweaver H, Burk D: The determination of enzyme dissociation constants. J Am Chem Soc 1934;56:658–666.

21   Gishan FK, Wilson FA: Developmental maturation of D-glucose transport by rat jejunal brush border membrane vesicles. Am J Physiol 1985;248:G87–G92.

22   Kabata H, Inui K-I, Itokawa Y: The binding of managanese to the brush-border membrane vesicles of the rat small intestine. Nutr Res 1989;9:791–799.

23   Kimberg D, Schachter D, Schenker H: Active transport of calcium by intestine, effects of dietary calcium. Am J Physiol 1961;200:1256–1262.

24   Gishan FK, Arab N, Nylander W: Characterization of calcium uptake by brush border membrane vesicles of human small intestine. Gastroenterology 1989;96:122–129.

25   Drüeke T, Hennessen U, Nabarra B, Ben Nasr L, Dang P, Thomasset M, Lacour B, Coudrier E, McCarron DA: Ultrastructural and functional abnormalities of the intestinal and renal epithelium in the spontaneously hypertensive rat. Kidney Internat 1990;37:1438–1448.

26   Emmer E, Rood RP, Wesolek JH, Cohen ME, Braithewaite RS, Sharp GWG, Murer H, Donowitz M: Role of calcium and calmodulin in the regulation of the rabbit ileal brush-border membrane Na^+/H^+ antiporter. J Membr Biol 1989;108:207–215.

27   Brushi G, Brushi ME, Caroppo M, Orlandini G, Pavarani C, Cavatorta A: Cytoplasmic free $[Ca^{2+}]$ is increased in platelets of spontaneously hypertensive rats and essential hypertensive patients. Clin Sci 1985;68:179–184.

28   Resnick LM, Gupta RK, Bhargava KK, Laragh JH: RBC cytosolic free calcium levels in hypertension. Relation to blood pressure and other cations. Am J Hypertens 1990;3:59A.

29   Vincenzi FF, Linder A, Hinds TR: Elevated intracellular free calcium in the red blood cells of human hypertensives: signal of change in the calcium pump and leak system. Am J Hypertens 1990;3:52A.

Uta Hennessen, Inserm U 90, Hôpital Necker, 161, rue de Sèvres,
F–75743 Paris Cedex 15 (France)

Morii H (ed): Calcium-Regulating Hormones. I. Role in Disease and Aging.
Contrib Nephrol. Basel, Karger, 1991, vol 90, pp 49–53

Mechanism of Hypercalciuria in Essential Hypertension and in Primary Nephrolithiasis

Carmine Zoccali, Francesca Mallamaci, Giuseppe Curatola,
Fiorella Cuzzola, Maurizio Postorino, Saverio Parlongo,
Filippa Salnitro, Adalgisa Curatola

Divisione di Nefrologia e Centro di Fisiologia Clinica, Reggio Calabria, Italy

Hypercalciuria has been consistently found in various strains of hypertensive rats [1, 2] as well as in patients with essential hypertension [3–7]. The pathogenesis of this alteration is very much debated because it has been attributed to disparate causes such as high salt intake [7], intestinal hyperabsorption of calcium [5] and urinary calcium leak [3, 4, 6]. Furthermore, it is still unclear whether the mechanism(s) responsible for enhanced urinary calcium excretion in essential hypertension differs from that in nephrolithiasis. Since previous studies were performed in unselected essential hypertensives on a free diet, in the present study we elected to compare in controlled metabolic conditions hypercalciuric hypertensives with normotensive hypercalciuric stone formers and with normotensive healthy subjects.

Methods

Subjects. Fifteen patients with essential hypertension and 10 normotensive hypercalciuric stone formers selected from the population attending our outpatient clinic participated in the study. These patients displayed normal renal function and frank hypercalciuria (i.e. CaU exceeding 6.25 mmol/day in at least two urinary collections on free diet). The third group was composed of 11 normotensive healthy subjects recruited from the medical and nursing staffs. The three groups were well matched for age (hypercalciuric hypertensives: 41 years (range 28–63); hypercalciuric stone formers: 38 (20–50); healthy subjects: 37 (26–47)), sex (M10/F5, 6/4 and 6/4, respectively) and body weight (74 kg (59–93), 70 (53–90) and 71 (57–92)). No patient had ingested any medication for at least 1 month before the study. Six hypertensive patients had passed urinary stones and 2 of them had evidence of nonobstructive nephrolithiasis. None of the normotensive stone formers had urinary obstruction.

Protocol. The protocol was in conformity with the ethical guidelines of our institution. Each participant was studied while consuming a diet with restricted calcium (10 mmol/day for 5 days) and again after 5 days of high calcium intake (35 mmol/day). High calcium intake was achieved by supplementing the calcium-restricted diet with two tablets of Calcium Forte Sandoz (each containing 12.5 mmol of elemental calcium in the form of calcium gluconate and calcium galactobionate). Sodium intake was fixed at 150 mmol/day throughout the study. The order of administration of the two diets was fully randomized and balanced. All subjects were studied as outpatients and were instructed to maintain their normal activities although vigorous exercise was not allowed. On the fourth and fifth days of each diet period they performed two consecutive 24-hour urine collections. On the fifth day, between 8 and 9 a.m., each participant was admitted to the research center under fasting conditions. Blood sampling was performed by an i.v. cannula after 40 min of supine rest.

Analytical Methods. Serum and urinary electrolytes measurements were performed in the routine clinical chemistry laboratory. Serum intact PTH was measured using the Allegro immunoradiometric assay (Nichols Institute, San Juan, Capistrano, Calif.). All samples from an individual subject were analyzed in duplicate in the same assay.

Statistical Analysis. For urinary electrolytes the average value of the two urinary collections was considered for statistical analysis. Within and between groups comparisons were performed by the paired and unpaired t tests as appropriate.

Results

Urine and Serum Electrolytes

As shown in figure 1, at high calcium intake urinary calcium in essential hypertensives was almost identical to that in stone formers and much higher ($p < 0.01$) than in healthy subjects. At low calcium intake, calcium excretion fell by about 20% in essential hypertensives. Such a fall was significantly ($p < 0.01$) less than that in stone formers (50%). However, also stone formers too were unable to reduce calcium excretion to the level attained by healthy subjects ($p < 0.01$). These differences were independent of sodium intake in that urinary sodium was identical in the three groups at low calcium and showed a similar rise with high calcium (table 1). In either diet period, the urinary phosphate and magnesium in essential hypertensives tended to be higher than in healthy controls (NS) but significantly less ($p < 0.01$) than in stone formers. Both essential hypertensives and stone formers displayed raised ($p < 0.01$) urinary potassium at high calcium intake. At low calcium intake, however, potassium excretion fell to normal in the first but not in the second group (table 1).

Serum total calcium fell significantly ($p < 0.01$) with low calcium in hypercalciuric hypertensives while it remained unchanged in the other groups (fig. 1). Serum phosphate and magnesium tended to be less in essential hypertensives and in stone formers than in healthy subjects in both diet periods (table 1).

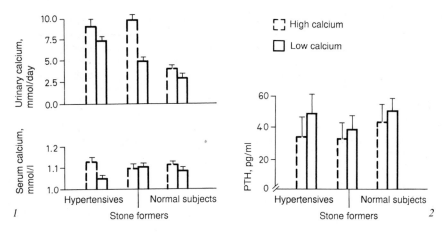

Fig. 1. Urinary and serum calcium in hypercalciuric hypertensives, in hypercalciuric stone formers, and in normal subjects.

Fig. 2. Serum PTH in the three study groups.

Serum PTH

Serum PTH in essential hypertensives did not differ from that in healthy subjects at low or at high calcium intake. PTH in stone formers was similar to that in the other groups at high calcium intake but was significantly reduced at low calcium intake (p < 0.02) (fig. 2).

Discussion

In the last 10 years hypercalciuria has emerged as a very frequent biochemical alteration in essential hypertension. McCarron et al. [3] and Strazzullo et al. [4] were the first to suggest that this type of disturbance depends on urinary calcium leak and that it is accompanied by mild secondary hyperparathyroidism. Other investigators contended that hypercalciuria is sustained by intestinal hyperabsorption of calcium secondary to renal phosphate leak [5] or that it merely reflects a high sodium intake [7]. We have found that under controlled metabolic conditions hypercalciuric hypertensives are unable to reduce appropriately calcium excretion when the intake of the cation is low and that they develop mild hypocalcemia. Such a response pattern clearly indicates that this form of hypercalciuria is mainly sustained by a renal leak. It is interesting to note that, in keeping with a study by Coe et al. [8], hypercalciuric stone formers manifested a substantial fall in calcium excretion without any change in serum calcium.

Table 1. Serum and urine electrolytes in the three study groups

	Essential hypertensives		Stone formers		Healthy subjects	
	high	low	high	low	high	low
Urine (24 h)						
Sodium, mmol	190 (63)	152 (50)	193 (52)	168 (70)	182 (55)	158 (39)
Phosphate, mmol	19.2 (7.5)	23.0 (6.7)	27.3 (5.7)	29.6 (9.5)	16.7 (5.1)	20.7 (5.1)
Magnesium, mmol	3.4 (1.1)	3.9 (1.5)	4.1 (1.4)	4.5 (1.5)	2.9 (0.8)	3.2 (0.8)
Potassium, mmol	66 (22)	56 (19)	74 (15)	76 (17)	57 (18)	57 (18)
Serum						
Sodium, mmol	140 (2)	139 (3)	139 (3)	139 (3)	141 (2)	139 (3)
Phosphate, mmol	1.00 (0.16)	0.90 (0.25)	1.00 (0.22)	0.90 (0.22)	1.01 (0.14)	1.01 (0.12)
Magnesium, mmol	0.71 (0.08)	0.71 (0.10)	0.72 (0.05)	0.68 (0.07)	0.75 (0.09)	0.75 (0.08)
Potassium, mmol	3.6 (0.3)	3.6 (0.3)	3.8 (0.3)	4.1 (0.4)	3.7 (0.4)	3.8 (0.3)

High and low refer to high or low calcium intake. Values in brackets are standard deviations.

The mechanism responsible for hypercalciuria in essential hypertension, therefore, does not coincide with that in nephrolithiasis.

Slight PTH elevations in essential hypertensives were reported in the studies of McCarron et al. [3] and Strazzullo et al. [4] and interpreted as a physiologic response to enhanced urinary calcium excretion. Other investigators, however, failed to confirm this finding [5–7]. In the present study we have found that at low calcium intake serum PTH in hypercalciuric hypertensives is similar to that in normotensive healthy subjects. This may entail a defective rather than a physiologic response in that serum calcium fell more markedly in essential hypertensives. Hypercalciuria in essential hypertension, therefore, may at least in part depend on inappropriate PTH response to the calcium leak. Similar observations have recently been made by Young et al. [9] who also noted a tendency for lower serum $1,25(OH)_2D_3$.

Summary

We have studied the metabolic response to changes in calcium in 15 hypercalciuric essential hypertensives, in 8 normotensive hypercalciuric stone formers and in 11 normotensive healthy subjects matched for age and sex. At variance with hypercalciuric stone formers, at low calcium intake hypercalciuric hypertensives did not appropriately reduce urinary calcium excretion and developed mild hypocalcemia. Furthermore, the PTH response to calcium deprivation was not appropriately enhanced in these patients. The data indicate that different mechanisms prevail in these two forms of hypercalciuria: the renal in essential hypertension and the intestinal in urolithiasis.

References

1   McCarron DA, Young NN, Ugoretz BA, Krutzik S: Disturbances of calcium metabolism in the spontaneously hypertensive rat. Hypertension 1981;3(suppl I):1162–1167.
2   Cirillo M, Galletti F, Strazzulo P, Torielli L, Melloni MC: On the pathogenetic mechanism of hypercalciuria in genetically hypertensive rats of the Milan strain. Am J Hypertens 1989;2:741–746.
3   McCarron DM, Pingree PA, Rubin RJ, Gaucher SM, Molitch M, Krutzik S: Enhanced parathyroid function in essential hypertension: A homeostatic response to a urinary calcium leak. Hypertension 1980;2:162–168.
4   Strazzullo P, Nunziata V, Cirillo M, Giannattasio R, Ferrara LA, Mattioli PL, Mancini P: Abnormalities of calcium metabolism in essential hypertension. Clin Sci 1983;65:137–141.
5   Montanaro D, Messa P, Messa M, Antonucci F, Paduano R, Mioni G: Evidenza di una accelerata cinetica del calcio e del fosforo nell'ipertensione arteriosa essenziale. Nefrologia, Dialisi, Trapianto. Milano, Wichtig, 1984, pp 87–90.
6   Zoccali C, Mallamaci F, Delfino D, Ciccarelli M, Parlongo S, Iellamo D, Moscato D, Maggiore Q: Phospate and divalent cations in essential hypertension. J Hypertens 1987;5(suppl 5):S323–S325.
7   Tillman DM, Semple PF: Calcium and magnesium in essential hypertension. Clin Sci 1988;75:395–402.
8   Coe FL, Favus JM, Crockett T, Strauss AL, Parks JH, Porat A, Gantt C, Sherwood LM: Effects of low calcium diet on urine calcium excretion, parathyroid function and serum $1,25(OH)_2D_3$ levels in patients with idiopathic hypercalciuria and in normal subjects. Am J Med 1982;72:25–31.
9   Young EW, Morris CD, McCarron DA: Increased urinary calcium excretion in essential hypertension. Am Soc Nephrol, 22nd Meeting, abstr 217A.

Dr. C. Zoccali, Centro di Fisiologia Clinica,
Via Sbarre Inferiori 39, I–89100 Reggio Calabria (Italy)

Morii H (ed): Calcium-Regulating Hormones. I. Role in Disease and Aging.
Contrib Nephrol. Basel, Karger, 1991, vol 90, pp 54–58

Effects of Dietary Calcium on Erythrocyte Sodium Ion Transport Systems in Spontaneously Hypertensive Rats

Kazutaka Fujito, Mamoru Yokomatsu, Nozomi Ishiguro,
Hiroo Numahata, Yasuhiko Tomino, Hikaru Koide

Division of Nephrology, Department of Medicine, Juntendo University School of
Medicine, Tokyo, Japan

An altered erythrocyte sodium ion transport of spontaneously hypertensive rats (SHR) has been observed in relation to the pathogenesis of hypertension [1, 2]. On the other hand, previous studies [3] indicated a salutary effect of a high calcium diet on blood pressure. However, the mechanism is obscure. The purpose of the present study was to determine the effects of dietary calcium intake on blood pressure and sodium ion transport systems of red blood cells in SHR.

Material and Methods

SHR and Wistar-Kyoto rats (WKY) aged 6 weeks were used for the experiments. Fifteen male SHR and 15 male WKY were divided into three groups, each of which was fed one of three levels of calcium, 0.1% (low Ca diet), 0.6% (normal Ca diet) and 4.0% (high Ca diet). Systolic blood pressure was measured weekly from 7 to 20 weeks of age using the tail-cuff method. At 20 weeks of age, blood samples were taken for the assay of intraerythrocyte sodium and potassium contents and sodium ion transport. The contents of sodium and potassium in the erythrocytes were measured by the method of Sachs and Welt [4]. Erythrocyte sodium efflux was measured by the method of Walter and Distler [5] with a slight modification. Red blood cells were preincubated with ^{22}Na for 2 h. After the washing, 1.0 ml of red blood cells was added to each of three different incubation mediums. During the incubation, red blood cells were serially taken, cooled and then centrifuged. The rate constant for sodium efflux was calculated from the slope of the regression line obtained by plotting the logarithm of ^{22}Na against incubation time. In this way, the ouabain-sensitive ^{22}Na efflux rate-constant (Na pump activity), the ouabain-insensitive bumetanide-sensitive ^{22}Na efflux rate constant (Na-K cotransport), and the ouabain-insensitive bumetanide-insensitive ^{22}Na efflux rate constant (Na permeability) were determined.

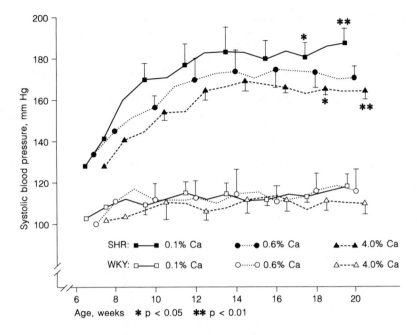

Fig. 1. Effect of dietary calcium on systolic blood pressure in SHR and WKY throughout the study period.

Results

Figure 1 shows the mean systolic blood pressure in SHR and WKY during the study period. At 18 weeks, there was a significant difference in blood pressure among the three groups of SHR. At 20 weeks, blood pressure was highest in the low Ca group of SHR, and lowest in the high Ca group. There was no significant difference in blood pressure among the three groups of WKY during the study period.

Table 1 shows the plasma levels of creatinine and electrolytes in SHR on the low, normal and high Ca diets, and in WKY on the normal Ca diet at 20 weeks of age. In SHR, the level of plasma creatinine was higher in the high Ca group than in the other groups. The level of total calcium was the highest in the high Ca group. Among SHR, the level of ionized calcium was the lowest in the low Ca group and highest in the high Ca group. With the normal Ca diet, ionized calcium was lower in SHR than WKY (p < 0.05). The levels of inorganic phosphate and magnesium were highest in the low Ca group and lowest in the high Ca group in SHR.

Table 1. Plasma creatinine and electrolytes in SHR and WKY at 20 weeks of age

Dietary calcium	SHR			WKY
	0.1%	0.6%	4.0%	0.6%
Creatinine, mg/dl	0.49 ± 0.06	0.46 ± 0.04	0.52 ± 0.04*	0.37 ± 0.06
Na, mEq/l	139.0 ± 1.6	138.3 ± 1.4	138.3 ± 1.4	142.0 ± 5.1
K, mEq/l	4.16 ± 0.46	4.22 ± 0.37	4.43 ± 0.34	4.18 ± 0.28
P, mg/dl	5.21 ± 0.61	4.84 ± 0.30	2.68 ± 0.22**	5.37 ± 0.38
Total Ca, mEq/l	5.12 ± 0.36	5.05 ± 0.21	5.58 ± 0.21**	4.94 ± 0.07
Ionized Ca, mEq/l	2.04 ± 0.24*	2.33 ± 0.07	2.59 ± 0.12**	2.43 ± 0.07*
Mg, mEq/l	1.53 ± 0.15*	1.34 ± 0.09	1.01 ± 0.14**	1.33 ± 0.07

* $p < 0.05$, ** $p < 0.01$ vs. 0.06% Ca SHR.

Figure 2 shows the contents of sodium and potassium in erythrocytes at 20 weeks in SHR and WKY. Among the SHR groups, sodium content was highest in the low Ca group and lowest in the high Ca group. Among the WKY groups, there was no significant difference in erythrocyte sodium content. SHR on the normal Ca diet had a higher erythrocyte sodium content than WKY ($p < 0.05$). There was no significant difference in erythrocyte potassium content between SHR and WKY.

Table 2 shows the ^{22}Na efflux rate constant at 20 weeks in SHR and WKY. The Na pump activity in SHR on the normal Ca diet was higher than that in WKY on the same diet. Among the SHR groups, Na pump activity was lowest in the low Ca group and highest in the high Ca group. In SHR, Na pump activity had an inverse correlation to the sodium content in erythrocytes ($r = -0.84$, $p < 0.01$) and a positive correlation to the level of ionized calcium in plasma ($r = 0.89$, $p < 0.01$). There was no significant difference between SHR and WKY in the Na-K cotransport. The Na permeability was higher in SHR than WKY on the normal Ca diet and there was no significant difference among the three SHR groups.

Discussion

In the present study, we attempted to determine whether dietary Ca intake affects blood pressure and cell membrane sodium transport in SHR. The low Ca group showed an enhancement of hypertension, while the high Ca group showed a suppression of the increase in blood pressure. The previous study [2] in SHR indicated that the high sodium content of erythrocytes was due to the increased permeability to sodium. The

Table 2. Constants of erythrocyte ^{22}Na efflux rate in SHR and WKY at 20 weeks of age

Dietary calcium	SHR			WKY
	0.1%	0.6%	4.0%	0.6%
Na$^+$-K$^+$ pump, h^{-1}	1.27 ± 0.14**	1.56 ± 0.10	1.85 ± 0.18*	1.36 ± 0.13*
Na$^+$-K$^+$ cotransport, h^{-1}	0.13 ± 0.04	0.09 ± 0.03	0.12 ± 0.06	0.14 ± 0.07
Na$^+$ permeability, h^{-1}	0.27 ± 0.04	0.30 ± 0.04	0.32 ± 0.03	0.20 ± 0.04**

* $p < 0.05$, ** $p < 0.01$ vs. 0.6% Ca SHR.

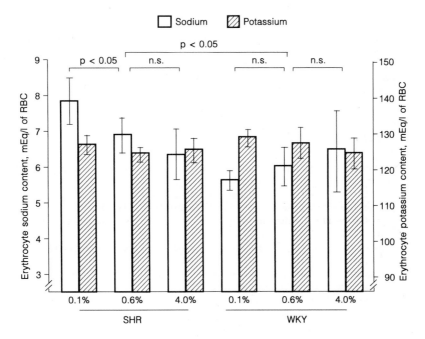

Fig. 2. Contents of sodium and potassium in the erythrocytes of SHR and WKY at 20 weeks of age.

increased Na pump activity might be a compensatory mechanism for the increased inward leakage of sodium. Our present results showed that dietary calcium intake influenced only Na pump activity, not Na permeability or Na-K cotransport. The changes in erythrocyte sodium content were due to changes in Na pump activity. Doris [6] suggested that high calcium intake reduces the digoxin-like factor in plasma. It is concluded that the effect of dietary calcium on blood pressure might be related to the alteration in sodium pump activity of cell membranes in SHR.

Summary

The alteration of sodium ion transport in red blood cells was observed in SHR and patients with essential hypertension. The purpose of the present study was to determine the effects of dietary calcium intake on blood pressure and sodium ion transport of red blood cells in SHR. The SHR were fed a diet with three different levels of calcium contents as follows: 0.1% (low), 0.6% (normal) and 4.0% (high) of calcium between 6 and 20 weeks of age. At 20 weeks of age, the levels of erythrocyte sodium efflux, sodium or potassium contents in the red blood cells were measured. On the high Ca diet, SHR showed an attenuation of the increase in blood pressure. On the low Ca diet, SHR showed an enhancement of hypertension. In proportion of increasing of dietary calcium contents, SHR had a lower level of sodium content in the RBC and a higher activity of the sodium pump. However, the passive sodium permeability and sodium-potassium cotransport in SHR were similar among the three different Ca diets. It is concluded that the amounts of dietary Ca might be related to the regulation of blood pressure by changing the sodium pump of the cell membrane in SHR.

References

1   Hardy FJ: Abnormalities of membrane transport in hypertension. Hypertension 1983;5:66–72.
2   Yokomatsu M: Erythrocyte sodium ion transport systems in DOCA-salt, Goldblatt and spontaneously hypertensive rats. Jap J Nephrol 1986;23:19–28.
3   McCarron DA: Calcium in pathogenesis and therapy of human hypertension. Am J Med 1985;78:27–33.
4   Sachs JR, Welt LG: The concentration dependence of active potassium transport in the human red blood cell. J Clin Invest 1967;46:65–76.
5   Walter U, Distler A: Abnormal sodium efflux in erythrocytes of patients with essential hypertension. Hypertension 1982;4:205–210.
6   Doris PA: Digoxin-like immunoreactive factor in rat plasma. Effect of sodium and calcium intake. Life Sci 1988;42:783–790.

Kazutaka Fujito, MD, Division of Nephrology, Department of Medicine, Juntendo University School of Medicine, Hongo, Bunkyo-ku, Tokyo 113 (Japan)

Morii H (ed): Calcium-Regulating Hormones. I. Role in Disease and Aging.
Contrib Nephrol. Basel, Karger, 1991, vol 90, pp 59–64

Adaptation of Low-Phosphate Diet in Renal Brush Borders of Spontaneously Hypertensive Rats

Kiyoshi Hirano, Masaki Nagasawa, Kazuhiro Saito, Yasuhiko Tomino, Hikaru Koide

Division of Nephrology, Department of Medicine, Juntendo University School of Medicine, Tokyo, Japan

The brush-border membrane (BBM) of proximal tubules, when isolated from the cell, retains its in vivo capacity for inorganic phosphate (P_i) transport. There is a close relationship between changes in the tubular P_i transport system by BBM vesicles under various experimental conditions [1]. This relationship suggests that the Na^+-dependent P_i transport system in proximal BBM vesicles is essential in the regulation of overall tubular P_i reabsorption.

Na^+-dependent P_i transport is mainly regulated by an intrinsic ability to adapt to the ambient P_i concentration [2, 3]. A low-phosphate diet (LPD) increases P_i uptake by BBM vesicles, and high-phosphate diet has the opposite effect [2, 3].

Disturbances in phosphate metabolism have been found in spontaneously hypertensive rats (SHR) [4, 5] and in hypertensive human subjects. Recent studies have demonstrated hypophosphatemia and hypophosphaturia in SHR.

The purpose of this study was to study the adaptation to a LPD in renal BBM vesicles of the superficial or deep cortex of kidneys in SHR. Wistar-Kyoto (WKY) rats were used as controls.

Materials and Methods

The experiments were performed on male 8-week-old SHR and WKY rats kept under standard conditions. They were fed a commercial chow containing 1.17% Ca, 200 IU/100 g of vitamin D, and either 0.17% (for the LPD) or 0.49% P (for the diet with a normal level of phosphate: NPD), for 6 days. The rats had free access to distilled water. They were anesthetized with pentobarbital sodium (40 mg/kg body weight). A catheter was placed in the femoral vein, for intravenous perfusion and blood sampling, and another was placed in the bladder for urine collection.

Table 1. Serum phosphate, calcium, and clearance of these substances

	Serum P_i	Serum Ca	Serum Mg	C_{cr}	FEP_i	FECa	FEMg
SHR							
LPD	4.85 ± 0.23	9.76 ± 0.24	2.40 ± 0.11	0.71 ± 0.03	0.27 ± 0.05	2.70 ± 0.48	16.7 ± 3.35
NPD	7.07 ± 0.36	9.48 ± 0.19	2.72 ± 0.15	0.76 ± 0.03	13.1 ± 0.99	0.29 ± 0.03	1.12 ± 0.43
WKY							
LPD	5.56 ± 0.43	9.24 ± 0.37	2.55 ± 0.12	0.78 ± 0.07	0.03 ± 0.01	4.08 ± 1.50	11.5 ± 1.95
NPD	7.53 ± 0.32	9.13 ± 0.11	2.73 ± 0.12	0.79 ± 0.05	9.95 ± 0.94	0.45 ± 0.09	3.72 ± 1.84
SHR LPD vs. NPD	$p < 0.01$	NS	NS	NS	$p < 0.01$	$p < 0.01$	$p < 0.01$
WKY LPD vs. NPD	$p < 0.01$	NS	NS	NS	$p < 0.01$	$p < 0.01$	$p < 0.05$
LPD SHR vs. WKY	NS	NS	NS	NS	$p < 0.01$	NS	NS
NPD SHR vs. WKY	NS	NS	NS	NS	$p < 0.05$	NS	NS

Results are mean $\pm$ SE. NS = Not significant. There were 10 rats in each group.

In a 30-min period, samples were obtained and urine and serum samples were assayed for calculation of the clearance of creatinine, P_i, Ca, and Mg; immediately afterwards, the rats were killed. The in vivo fractional excretion of P_i (FEP_i), calcium (FECa), and magnesium (FEMg) was calculated from differences between the filtered load of P_i, Ca, and Mg and the absolute amounts of these substances excreted in the urine.

BBM vesicles were prepared from superficial and deep cortical tissue of the rat kidney. 0.3-mm thick slices were cut from the outer surface of the kidneys with a sharp scalpel. The kidneys were then sectioned transversely, medullary tissue was discarded, and 0.3-mm thick slices were obtained. BBM vesicles were isolated by $MgCl_2$ precipitation and purified by chromatography on a glass beads column with pores of defined size.

The concentrations of enzymes of the final suspension compared with that of crude homogenate were not increased in markers of mitochondria, endoplasmic reticulum, and the basolateral membrane. The activity of alkaline phosphatase (AL-P), a marker of the BBM, was about 20-fold that of the crude homogenate of cortex. The activities of Al-P and two other marker enzymes of the BBM, leucine aminopeptidase (LAP) and γ-glutamyl transpeptidase (γ-GTP) were assayed in suspensions of BBM vesicles prepared from superficial or deep cortical tissue.

Transport studies were done on the same day as the preparation of the BBM vesicles. Uptake of labeled substrates by the isolated vesicles was measured by a rapid-filtration technique with the use of a 0.45-μm filter (Millipore Corp., Bedford, Mass.). The amount of labeled substrate remaining on the filter was measured by liquid scintillation counting. The optimum conditions for the uptake studies were found in preliminary experiments. The apparent K_m and V_{max} were calculated from regression lines obtained from least-squares analysis in Eadie-Hofstee plots. The statistical significance of differences was evaluated by the Student's t test.

Results

The effects of the LPD on the serum concentrations of P_i, Ca, and Mg are shown in table 1, together with the clearance of these substances and of creatinine. The LPD caused a significant decrease in serum P_i in SHR and WKY rats. The clearance of creatinine did not differ significantly among the four groups. FEP_i decreased significantly more in WKY rats in response to the LPD than in SHR. The LPD caused a significant increase in Al-P activity, but had no effect on the activities of LAP and γ-GTP.

The course of in vitro P_i uptake by BBM vesicles was examined. In the presence of a 100-mM sodium gradient, the course of P_i uptake by BBM vesicles overshot the curve. The absolute uptake of P_i by BBM vesicles at 60 min, the time of equilibrium, was not significantly different among the four groups in either the superficial cortex or the deep cortex. This indicated that BBM vesicles obtained from each group in both parts of the cortex were comparable in size. The rate of Na^+-independent P_i uptake (with KSCN in medium instead of NaSCN) showed no overshoot, and did not differ in BBM vesicles from the different groups and parts of the cortex. The P_i concentration of the incubation medium varied from 12.5 to

Table 2. Kinetic parameters of phosphate transport by BBM vesicles

	Superficial cortex		Deep cortex	
	V_{max} (nmol mg^{-1} protein)	K_m (μM)	V_{max} (nmol mg^{-1} protein)	K_m (μM)
SHR				
LPD	4.52 ± 0.24 (NS)	47 ± 2.0	$6.14 \pm 0.21^*$	23 ± 4.8
NPD	4.53 ± 0.35 (NS)	51 ± 4.8	$3.86 \pm 0.01^*$	24 ± 0.3
WKY				
LPD	$10.9 \pm 0.20^{**}$	49 ± 3.0	4.30 ± 0.62 (NS)	24 ± 2.6
NPD	$5.23 \pm 1.18^{**}$	37 ± 0.9	2.69 ± 0.06 (NS)	36 ± 9.4

Results are mean $\pm$ SE. NS = Not significant; $^*p < 0.01$, $^{**}p < 0.05$. There were 5 BBM vesicle preparations in each group.

300 μM, and P_i uptake was measured after 20 s of incubation. P_i uptake reached saturation when the P_i concentration was increased in the medium.

The LPD had no effect on the rate of P_i uptake and V_{max} on superficial cortex of SHR, which showed that adaptation did not occur in BBM vesicles obtained from the superficial cortex of the SHR. In BBM vesicles from the deep cortex of these rats, the LPD increased the Na$^+$-gradient-dependent P_i uptake and the V_{max} of P_i uptake. In the BBM vesicles prepared from superficial cortex of WKY rat kidneys, P_i uptake and the V_{max} increased in response to the LPD, showing adaptation. P_i uptake and V_{max} both increased in deep cortical BBM vesicles from WKY rats on the LPD, but less than the superficial cortex.

Table 2 summarizes the kinetic parameters of P_i uptake by BBM vesicles. The apparent K_m did not differ significantly among the groups or between the superficial and deep cortex. In the SHR, there was no difference in the V_{max} of P_i uptake by superficial cortical BBM vesicles depending on diet. However, in the deep cortical vesicles of the BBM of these rats, the LPD increased the V_{max} significantly. In the WKY rats, the V_{max} of P_i uptake by superficial cortical BBM vesicles was increased significantly by the LPD.

Discussion

We did this study to find whether a change in the FEP$_i$ is related to P_i transport in superficial or deep cortical BBM vesicles.

The difference in the response to the LPD in terms of the FEP$_i$ of SHR and of WKY rats indicated a difference in P_i adaptation by the whole kidney between these strains. The adaptive response to the LPD in SHR was less than that in the WKY rats. P_i adaptation occurred in SHR via an increase in AL-P.

BBM vesicles prepared from the superficial cortex were presumably composed principally of S1 and S2 segments of the proximal tubule, and those from the deep cortex were principally S3 segments [6]. Evidence derived from saturation studies suggested that there are two P_i transporters dependent on a Na^+ gradient in the early proximal tubules, one with a high capacity and low affinity and the other with a low capacity and high affinity [6]. Up until now, it was thought that there is only one transporter in the late proximal tubule, one with a low capacity and high affinity. This serial arrangement of transporters would bring about reabsorption of most of the filtered P_i in the early proximal tubule and of a smaller amount in the late proximal tubule, where the largest concentration gradient is formed.

To judge from the results of kinetic analysis of our saturation studies, the affinity of the P_i transporter was not influenced by the LPD. On the other hand, the capacity of the P_i transporter in BBM vesicles prepared from the superficial cortex of WKY rat kidneys was significantly increased by the LPD. In the deep cortex of WKY rat kidneys, this capacity in the LPD group tended to be higher than that in the NPD group. In the SHR, the LPD had no effect on the capacity of the P_i transporter in BBM vesicles prepared from superficial cortex. However, the LPD increased this capacity significantly in the deep cortical BBM vesicles of SHR. Probably adaptation occurred mainly in the early proximal tubules in WKY rats, but adaptation occurred only in the late proximal tubules in SHR; none was observed in the early proximal tubules.

In conclusion, the adaptive response to the LPD in SHR was less than that in WKY rats. The adaptation seen in BBM vesicles from the superficial cortex of WKY rats was not seen in such vesicles obtained from SHR. The less adaptation in SHR suggests a defect in P_i transport in superficial BBM vesicles of these rats.

Summary

A LPD seems to increase the P_i uptake by vesicles of the BBM of the renal cortex. This study was done to find if there was adaptation to a LPD in the BBM vesicles from the superficial or deep cortex in SHR, with WKY rats as control. The fractional excretion of P_i of SHR on the LPD was higher than that of WKY rats ($p < 0.01$). The V_{max} of P_i uptake in BBM vesicles from superficial cortex of WKY rats on the LPD was greater ($p < 0.05$) than in such rats on a diet with a normal level of phosphate. Thus, the adaptation to a LPD was normal in WKY rats. However, in BBM vesicles from the superficial cortex of SHR kidneys, the difference in this V_{max} depending on diet was insignificant. In BBM vesicles of the deep cortex of SHR kidneys, this V_{max} was higher (p < 0.01) on the LPD. The apparent K_m was not significantly different in different groups or parts of the renal cortex. These results suggest that BBM vesicles in the superficial cortex of SHR kidneys did not adapt to the LPD. The less adaptation in SHR in vivo indicates that there may be a defect in P_i transport in BBM vesicles of the superficial cortex of SHR kidneys.

References

1   Hammerman MR: Phosphate transport across renal proximal tubular cell membranes. Am J Physiol 1986;251:F385–F398.
2   Troehler U, Bonjour JP, Fleisch H: Inorganic phosphate homeostasis. Renal adaptation to the dietary intake in intact and thyroparathyroidectomized rats. J Clin Invest 1976;57:264–273.
3   Quamme G, Biber J, Murer H: Sodium-phosphate cotransport in OK cells: Inhibition by PTH and 'adaptation' to low phosphate. Am J Physiol 1989;257:F967–F973.
4   Hirano K: Transport system of renal brush border membrane in spontaneously hypertensive rats. Jpn J Nephrol 1988;30:1–9.
5   Bindels RJM, Geertsen JAM, Van Os CH: Increased transport of inorganic phosphate in renal brush borders of spontaneously hypertensive rats. Am J Physiol 1986;250:F470–F475.
6   Walker JJ, Yan TS, Quamme GA: Presence of multiple sodium-dependent phosphate transport processes in proximal brush-border membranes. Am J Physiol 1987;252:F226–F231.

Dr. K. Hirano, Division of Nephrology, Department of Medicine, Juntendo University School of Medicine, Hongo 2-1-1 Bunkyo-ku, Tokyo 113 (Japan)

Role of Calcium-Regulating Hormones in the Relationship between Calcium and Hypertension

Morii H (ed): Calcium-Regulating Hormones. I. Role in Disease and Aging.
Contrib Nephrol. Basel, Karger, 1991, vol 90, pp 65–71

Parathyroid Glands and Cardiovascular Functions

Peter K.T. Pang, Richard Z. Lewanczuk, Christina G. Benishin

Department of Physiology, University of Alberta, Edmonton, Alta., Canada

It has been known for a long time that hyperparathyroid patients are often hypertensive [1–3]. The reason for the correlation between parathyroid function and the clinical observation of hypertension is obscure. Specifically, it is not well understood why some hyperparathyroid patients are hypertensive. There are recent speculations that parathyroid hormone, PTH, may be the causative factor in these patients. In the present paper, evidence for and against PTH being the cause of hypertension is presented. Some of our recent studies suggest that the parathyroid gland may secrete a new hypertensive factor which is distinct from PTH. These findings will be summarized in this paper.

Evidence for PTH as the Causative Factor in Some Forms of Hypertension

There are many observations which led to the speculation that PTH might be the cause of some forms of hypertension. First, recent studies have shown that, in spontaneously hypertensive rats (SHR), the plasma PTH level is often elevated [4, 5]. Such a correlation between plasma PTH levels and hypertension has also been reported in human hypertension [6–9]. Second, in a recent publication, the parathyroid gland of SHR was shown to be enlarged [10]. Of course, it has been reported many times, and it is a known general clinical phenomenon that hyperparathyroid patients are often hypertensive, although not always so [1–3]. Third, surgical removal of the hyperactive parathyroid gland often, but not always, returns the blood pressure of such hypertensive patients to normal [11–14]. Parathyroidectomy of SHR has been shown to reduce blood pressure [15–17]. Fourth, relative hypocalcemia has been correlated with essential hypertension in some hypertensive patients [18–20]. It is well known that hypocalcemia can stimulate the secretion of PTH and thus produce an abnormally high plasma level of PTH. As previously noted, such increased PTH levels have been observed in SHR [4, 5]. Fifth, a high-calcium diet can reduce plasma PTH levels and blood pressure in SHR. In humans, calcium supplementation can also lower blood pressure in some patients [21–24], although in some studies, such a beneficial effect of high calcium intake cannot be demonstrated [25–27]. Sixth, PTH has been shown to increase the calcium content of some tissues [28–31], and increased intracellular free calcium concentrations have been reported in tissues of hypertensive patients and animals [32–35]. All these observations taken together provide some very compelling evidence that, indeed, PTH may be the cause of some forms of hypertension in humans and SHR.

Evidence against PTH as the Causative Factor in Some Forms of Hypertension

Although the evidence for PTH as the cause of hypertension is substantial, there are also strong arguments that indicate that PTH is not. First, not all hyperparathyroid patients are hypertensive, even though the plasma PTH levels of such patients are, by definition also high. Second, parathyroidectomy does not always correct the hypertension. In fact, in most cases, surgical operation resulted in an increase in blood pressure in a small proportion of patients [1, 12–14]. Third, in hyperparathyroid hypertensive patients, the plasma calcium concentration is above normal, although it can be argued that the hypercalcemia is secondary to the increase in plasma PTH levels. It will then be difficult to explain why, in hypertensives, relative hypocalcemia is persistent in spite of the high plasma PTH level. Fourth, parathyroid hormone has been repeatedly reported to have a hypotensive effect when injected into mammalian, as well as many nonmammalian, species [36–38]. This is a specific property of the molecule, as various synthetic preparations of active human, bovine and rat PTH fragments have the same hypotensive effect [39]. The part of the molecule responsible for the vascular action has been shown to be different from that for the hypercalcemic action [40]. The 24–28 residues are probably the active part of the molecule responsible for the vascular action [41]. PTH specifically dilates the coronary and renal vascular beds [42, 43], suggesting that the vascular effect is not a nonspecific pharmacological phenomenon. It is, therefore, difficult to envisage how a hypotensive hormone can be the cause of hypertension. Fifth, PTH stimulates vascular adenylate cyclase and increases cAMP levels in blood vessels [44, 45] and cAMP is hypotensive. Sixth, in the study of the mechanism of its vasodilating action, PTH has been reported to decrease calcium uptake by rat tail artery. This correlates well with the relaxing effect of PTH in this tissue [46]. In a recent series of experiments, vascular smooth muscle cells from rat tail arteries were cultured and their calcium channels studied. Two types of calcium channels, L and T types, were identified [47]. PTH specifically inhibited the L, but not the T, channel. This effect was dose related and could be reversed by BAY-K-8644, a synthetic L channel opener. Furthermore, a denatured and vascularly inactive preparation of PTH had no effect on the L channel. The calcium channel effect of PTH was blocked by the antagonistic fragment, bPTH (3–34). PTH behaves like an endogeneous calcium channel blocker. It is difficult to believe that a calcium channel blocker can be the cause of hypertension. However, it has been reported that PTH can increase calcium content in some tissues [28–31]. It is possible that PTH may have a paradoxical effect on calcium channels in different tissues. In our calcium channel studies, the effect of PTH on cultured rat cardiac myocytes was also tested. PTH amplified the L channel, and such amplification was blocked by nifedipine, a calcium channel blocker, and by bPTH (3–34) [Wang, Karpinski and Pang, this meeting]. These data indicate that PTH, indeed, can have opposite effects on cellular calcium balance. However, as it inhibits calcium entry in vascular smooth muscle, which is directly related to blood pressure, its overall effect is vasodilation and not hypertension.

Parathyroid Origin of PHF: A New Circulating Hypertensive Factor

Since PTH cannot be the causative factor for hypertension, an explanation must be provided for the role of the parathyroid gland in hypertension. We recently reported the discovery of a circulating hypertensive factor which produced delayed hypertension in normotensive assay rats. Such a delayed action is unique, as almost all known vasoconstrictors have immediate actions [48]. This factor is parathyroid in origin, as parathyroidectomy reduced blood pressure in SHR with a concurrent disappearance of the factor from the plasma. Transplantation of parathyroid glands from SHR into normotensive rats produced hypertension with the appearance of the factor in the plasma [49]. The parathyroid origin of the factor was further demonstrated by the secretion of the factor by SHR parathyroid glands in culture [Benishin, Lewanczuk and Pang, unpublished observations]. In a human hyperparathyroid study, the presence of the factor in plasma was correlated with hypertension. Parathyroidectomy on these factor-positive patients led to a reduction in blood pressure and the disappearance of the factor from the plasma [Lewanczuk and Pang, unpubl. observations]. We named this factor PHF – parathyroid hypertensive factor. PHF is probably secreted by a special cell type found in the parathyroid gland of SHR. The percentage of these cells in the whole gland correlated significantly with the blood pressure and the amount of PHF in the plasma of SHR rats [50]. Furthermore, PHF increased calcium uptake into rat tail artery in vitro and the peak uptake was also delayed, as in the case of the blood pressure effect [48]. In cultured vascular smooth muscle cells from rat tail arteries, PHF increased intracellular free calcium levels [Shan and Pang, unpubl. observations]. Feeding SHR rats with low calcium food resulted in high levels of PHF, while in rats fed a high-calcium diet, PHF was absent from the plasma and the blood pressure was significantly lower than that of rats fed a low-calcium diet [51]. PHF cannot be PTH, because in the PHF assay, PTH produced a decrease, rather than an increase, in blood pressure. All these data provide an explanation for the observations which led to the suggestion that PTH might be the causative factor in some forms of hypertension. It seems that PHF may be such a factor.

Conclusion

The parathyroid gland is involved in some forms of hypertension, but PTH is not the hormone responsible. PHF, the recently discovered circulating hypertensive factor, may be the causative factor in some forms of hypertension. It is secreted by a novel cell type in the parathyroid gland. It increases calcium uptake and intracellular free calcium in vascular smooth muscle. Its secretion is inhibited by a high-calcium diet. The existence of PHF may thus explain the involvement of parathyroid glands in hypertension.

References

1   Hellstrom J, Birke G, Edvall CA: Hypertension in hyperparathyroidism. Br J Urol 1958;30:13–24.
2   Rosenthal FD, Roy S: Hypertension and hyperparathyroidism. Br Med J 1972;4:396–397.
3   Leug MC: Hypertension and primary hyperparathyroidism: A five year case review. South Med J 1982;75:1371–1374.
4   McCarron DA, Yung NN, Ugoretz BA, Krutzik S: Disturbances of calcium metabolism in the spontaneously hypertensive rat. Hypertension 1981;I:162–167.
5   Stern N, Lee DBN, Silis V, Beck FWJ, Deftos L, Manolagas SC, Sowers JR: Effects of high calcium intake on blood pressure and calcium metabolism in young SHR. Hypertension 1984;6:639–646.

6   McCarron DA, Pingree P, Rubin RJ, Gaucher SM, Molitch M, Krutzik S: Enhanced parathyroid function in essential hypertension: A homeostatic response to a urinary calcium leak. Hypertension 1980;2:162–168.

7   Strazzullo P, Nanziata V, Cirillo M, Giannattasio R, Ferrara LA, Mattiolli PL, Mancini M: Abnormalities of calcium metabolism in essential hypertension. Clin Sci 1983;65:137–141.

8   Hvarfner A, Bergström R, Mörlin C, Wide L, Ljunghall S: Relationships between calcium metabolic indices and blood pressure in patients with essential hypertension as compared with a healthy population. J Hypertens 1987;5:451–456.

9   Grobbee DE, Hackeng WHL, Birkenhager JC, Hofman A: Raised plasma intact parathyroid hormone concentrations in young people with mildly raised blood pressure. Br Med J 1988;296:814–816.

10   Merke J, Lucas PA, Szabo A, Cournot-Witmer G, Mall G, Bouillon R, Drüeke, T, Mann J, Ritz E: Hyperparathyroidism and abnormal calcitriol metabolism in the spontaneously hypertensive rat. Hypertension 1989;13:233–242.

11   Madhaven T, Frame B, Block MA: Influence of surgical correction of primary hyper-parathyroidism on associated hypertension. Arch Surg 1970;100:212–214.

12   Diamond TW, Botha JR, Wing J, Meyers AM: Parathyroid hypertension: a reversible disorder. Arch Intern Med 1986;146:1709–1712.

13   Niederle B, Roka R,Woloszczyk W, Klaushofer K, Kovarik J, Schernthaner G: Success-ful parathyroidectomy in primary hyperparathyroidism: A clinical follow-up study of 212 consecutive patients. Surgery 1987;102:903–909.

14   Rapado A: Arterial hypertension and primary hyperparathyroidism. Am J Nephrol 1986;6(suppl. 1):49–50.

15   Mann JFE, Wiecek A, Bommer J, Ganten U, Ritz E: Effects of parathyroidectomy on blood pressure in spontaneously hypertensive rats. Nephron 1987;45:46–52.

16   Schlieffer R, Berthelot A, Pernot F, Gairard A: Parathyroids, thyroid and development of hypertension in SHR. Jpn Circ J 1981;45:1272–1279.

17   Baksi SN: Hypotensive action of parathyroid hormone in hypoparathyroid and hyper-parathyroid rats. Hypertension 1988;11:509–513.

18   McCarron DA: Low serum concentrations of ionized calcium in patients with hyperten-sion. N Engl J Med 1982;307:226–228.

19   Resnick LM, Laragh JH, Sealy, JE, Alderman MA: Divalent cations in essential hypertension. Relations between serum ionized calcium, magnesium and plasma renin activity. N Engl J Med 1983;309:888–891.

20   Folsom A, Smith C, Prineas R, Grimm R: Serum calcium fractions in essential hypertension and matched normotensive subjects. Hypertension 1986;8:11–15.

21   Belizan JM, Villar J, Pineda O, González AE, Sainz E, Garrera G, Sibrian R: Reduction of blood pressure with calcium supplementation in young adults. JAMA 1983;249:1161–1165.

22   Johnson NE, Smith ILM, Freudenheim JL: Effects on blood pressure of calcium supplementation of women. Am J Clin Nutr 1985;42:12–17.

23   McCarron DA, Morris CD: Blood pressure response to oral calcium in persons with mild to moderate hypertension. Ann Intern Med 1985;103:825–831.

24   Strazzullo P, Siano A, Guglielini S, DiCarlo A, Galletti F, Cirillo M, Mancin M: Controlled trial of long term oral calcium supplementation in essential hypertension. Hypertension 1986;8:1084–1088.

25   Capuccio FP, Markandu ND, Singer DRJ, Smith SJ, Shore AC, MacGregor FA: Does oral calcium supplementation lower high blood pressure? A double blind study. J Hypertens 1987;5:67–71.

26 Siani A, Strazzullo P, Guglielni S, Pacioni D, Giacco A, Iacone R, Mancini M: Controlled trial of low calcium versus high calcium intake in mild hypertension. J Hypertens 1988;6:253–256.

27 Zoccali C, Mallamaci F, Delfino D, Ciccarelli M, Parlongo S, Iellamo D, Moscato D, Maggiore Q: Double blind randomized crossover trial of calcium supplementation in essential hypertension. J Hypertens 1988;6:451–455.

28 Bogin E, Massry SG, Harary I: Effect of parathyroid hormone on heart cells. J Clin Invest 1981;67:1215–1227.

29 Zemel MB, Bedford BA, Standley PR, Sowers JR: Saline infusion causes rapid increase in parathyroid hormone and intracellular calcium levels. Am J Hypertens 1989;2:185–187.

30 Borle AB, Uchikawa T: Effects of parathyroid hormone on the distribution and transport of calcium in cultured kidney cells. Endocrinology 1978;102:1725–1732.

31 Goldstein DA, Massry SG: Effect of parathyroid hormone administration and its withdrawal on brain calcium and electroencephalogram. Miner Electrolyte Metab 1978;1:84–91.

32 Erne P, Bolli P, Burgisser E, Buhler FR: Correlation of platelet calcium with blood pressure. Effect of antihypertensive therapy. N Engl J Med 1984;310:1084–1088.

33 Bruschi G, Bruschi ME, Caroppo M, Orlandini G, Spaggiar M, Cavatorta A: Cytoplasmic free calcium is increased in the platelets of spontaneously hypertensive rats and essential hypertensive patients. Clin Sci 1985;68:179–184.

34 Furspan PB, Bohr DF: Calcium-related abnormalities in lymphocytes from genetically hypertensive rats. Hypertension 1986;8(suppl 2):I123–126.

35 Spieker C, Zidek W, von Bassewitz DB, Heck D: Age-dependent increase in arterial smooth muscle calcium in spontaneously hypertensive rats. Res Exp Med 1988;188:397–403.

36 Charbon GA: A diuretic and hypotensive action of a parathyroid extract. Acta Physiol Pharmacol Neerl 1966;14:52–53.

37 Berthelot A, Gairard A: Effet de la parathormone sur la pression artérielle et la contraction de l'aorte isolée de rat. Experientia 1975;31:457–458.

38 Pang PKT, Yang MCM, Oguro C, Phillips JG, Yee JA: Hypotensive actions of parathyroid hormone preparations in vertebrates. Gen Comp Endocrinol 1980;41:135–138.

39 Pang PKT, Tenner TE Jr, Yee JA, Yang M, Janssen HF: Hypotensive action of parathyroid hormone preparations on rats and dogs. Proc Natl Acad Sci USA 1980;77:675–678.

40 Pang PKT, Yang MCM, Keutmann HT, Kenny AD: Structure activity relationship of parathyroid hormone: Separation of the hypotensive and hypercalcemic properties. Endocrinology 1983;112:284–289.

41 Hong BS, Yang MCM, Liang JN, Pang PKT: Correlation of structural changes in parathyroid hormone with its vascular action. Peptides 1986;7:1131–1135.

42 Pang PKT, Janssen HF, Yee JA: Effects of synthetic parathyroid hormone on vascular beds of dogs. Pharmacology 1980;21:213–222.

43 Crass MF III, Pang PKT: Parathyroid hormone: A coronary artery vasodilator. Science 1980;207:1087–1089.

44 Nickols GA, Metz, MA, Cline WH Jr: Endothelium-independent linkage of parathyroid hormone receptors of rat vascular tissue with increased adenosine 3′,5′-monophosphate and relaxation of vascular smooth muscle. Endocrinology 1986;119:349–356.

45 Nickols GA: Increased cAMP in cultured vascular smooth muscle cells and relaxation of aortic strips by parathyroid hormone. Eur J Pharmacol 1985;116:137–144.

46 Pang PKT, Yang MCM, Sham JSK: Parathyroid hormone and calcium entry blockade in a vascular tissue. Life Sci 1988;42:1395–1400.

47 Pang PKT, Wang R, Shan J, Karpinski E, Benishin CG: Specific inhibition of L-type calcium channels by synthetic parathyroid hormone. Proc Natl Acad Sci USA 1990;87:623–627.

48 Lewanczuk RZ, Wang J, Zhang SR, Pang PKT: Effects of spontaneously hypertensive rat plasma on blood pressure and tail artery calcium uptake in normotensive rats. Am J Hypertens 1989;2:26–31.

49 Pang PKT, Lewanczuk RZ: Parathyroid origin of a new circulating hypertensive factor in spontaneously hypertensive rats. Am J Hypertens 1990;2:898–902.

50 Pang PKT, Benishin C, Kaneko T, Lewanczuk RZ: The relationship between the parathyroid gland and parathyroid hypertensive factor in spontaneously hypertensive rats. Am J Hypertens 1990;3:21A.

51 Lewanczuk RZ, Pang PKT: Effects of dietary calcium on the presence of a circulating hypertensive factor in spontaneously hypertensive rats. Am J Hypertens 1990;3:349–353.

Peter K.T. Pang, Division of Endocrinology, Department of Physiology,
University of Alberta, Edmonton, Alta., T6G 2H7 (Canada)

Morii H (ed): Calcium-Regulating Hormones. I. Role in Disease and Aging.
Contrib Nephrol. Basel, Karger, 1991, vol 90, pp 72–78

Effects of Parathyroid Hormone-Related Protein on Systemic and Regional Hemodynamics in Conscious Rats

A Comparison with Human Parathyroid Hormone

H. Kishimoto, K. Tsumura, S. Fujioka, S. Uchimoto, N. Yamashita, R. Suzuki, K. Yoshimaru, M. Shimura, O. Sasakawa, H. Morii

Second Department of Internal Medicine, Osaka City University Medical School, Osaka, Japan

Parathyroid hormone (PTH) is a major calcium-regulating hormone and there have been many reports of the effects of PTH on the cardiovascular system. In particular, a hypotensive effect of PTH has been shown in several animals [1, 2], which results from the direct relaxation of vascular smooth muscle by PTH [3]. Such PTH-induced vasodilation has been found in the vascular beds of the heart, liver, kidney and brain [4, 5]. A positive inotropic and chronotropic effect of PTH on the heart has been shown by in vitro and in vivo studies [6, 7].

PTH-related protein (PTHrp) was discovered as a humoral hypercalcemic factor of malignancy [8]. It has significant homology with PTH in the amino-terminal region and has been shown to bind the same receptor as PTH in rat bone cells and canine renal membranes [9]. However, there have been few reports about the effect of PTHrp on cardiovascular system [10, 11].

The purpose of this study is to investigate the global effect of human-PTH and PTHrp on systemic and regional hemodynamics in conscious rats.

Materials and Methods

All animals studied were male Sprague-Dawley rats weighing between 300 and 350 g. Polyethylene catheters (PE-50) were inserted into the femoral artery, the femoral vein, and the left ventricle via the right carotid artery under sodium pentobarbital anesthesia. Twenty-four hours later, in the conscious and unrestrained state, the mean arterial pressure (MAP) and heart rate (HR) were monitored continuously through the femoral artery catheter. After the

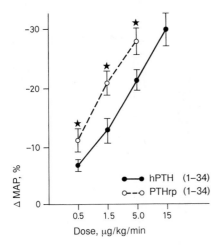

Fig. 1. The effect of hPTH and PTHrp on mean arterial pressure (MAP) in conscious unrestrained rats (n = 7). Values are the mean ± SEM. *p < 0.05, compared to preinjection.

stabilization of MAP and HR, all studies were performed. In the first study, the rats were given intravenous infusions of 0.5, 1.5, 5.0 and 15 μg/kg/min of human PTH-(1-34) (hPTH) or 0.5, 1.5, and 5.0 μg/kg/min of PTHrp-(1-34) via the femoral vein catheter. MAP and HR were monitored before and 5 min after the start of each peptide infusion. In the second study, systemic and regional hemodynamics were determined by the microsphere technique [12]. Radionuclide-labeled microspheres ([141]Ce and [95]Nb, New England Nuclear, USA) were suspended in saline before use. Then, 0.25 ml of the suspension of [141]Ce microspheres was injected into the left ventricle over a 15-second period, and 5 min after the intravenous infusion of hPTH (15 μg/kg/min) or PTHrp (5 μg/kg/min) the [95]Nb microspheres were injected. Blood withdrawal was begun at the rate of 0.67 ml/min through the femoral artery catheter, and was continued for 1 min. The radioactivity of all blood samples and organs were determined to calculate cardiac output and the blood flow of each organ was determined using a gamma scintillation counter.

The data in this paper are given as the mean ± SEM. Statistical significance of difference was evaluated using Student's paired and unpaired t tests. A p value of less than 0.05 was considered significant.

Results

PTHrp and hPTH decreased MAP in a dose-dependent manner with the hypotensive response to PTHrp being significantly greater than that to hPTH at each dose. For example, the dose of hPTH required to decrease MAP by 20% was about 4.0 μg/kg/min, whereas the dose of PTHrp required to decrease MAP by 20% was about 1.4 μg/kg/min. Therefore, PThrp was three times more potent than hPTH (fig. 1).

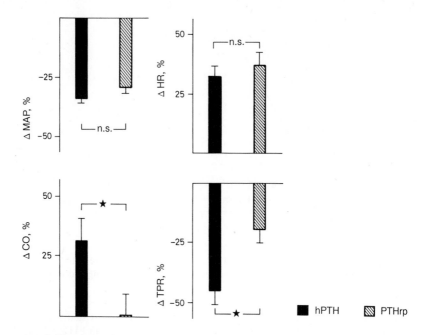

Fig. 2. The change in systemic hemodynamic parameters in response to hPTH (15 μg/kg/min, i.v.) and PTHrp (5 μg/kg/min, i.v.) in conscious rats (n = 7). MAP = Mean arterial pressure; HR = heart rate; CO = cardiac output; TPR = total peripheral resistance. Values are mean ± SEM. *p < 0.05, hPTH vs. PTHrp.

Table 1. Effect of hPTH or PTHrp on systemic hemodynamics in conscious rats

	hPTH (15 μg/kg/min), n = 7		PTHrp (5 μg/kg/min), n = 7	
	preinjection	postinjection	preinjection	postinjection
MAP, mm Hg	105 ± 3	72 ± 2**	104 ± 1	75 ± 3**
HR, beats/min	407 ± 15	539 ± 12**	345 ± 16	487 ± 23**
CO, ml/min	149 ± 8	195 ± 16*	132 ± 8	133 ± 17
TPR, mm Hg/ml/min	0.71 ± 0.04	0.39 ± 0.04**	0.80 ± 0.06	0.62 ± 0.07**

Values are the means ± SEM. MAP = Mean arterial pressure; HR = heat rate; CO = cardiac output; TPR = total peripheral resistance.
*p < 0.05, **p < 0.01, compared to preinjection values.

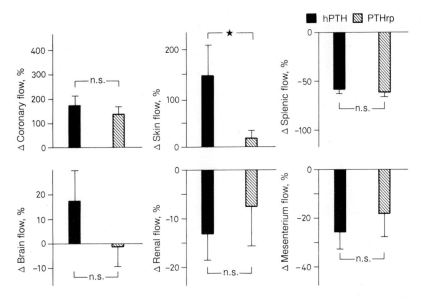

Fig. 3. The changes in regional blood flow in response to hPTH (15 μg/kg/min, i.v.) and PTHrp (5 μg/kg/min, i.v.) in conscious rats (n = 7). Values are the mean ± SEM. *p < 0.05, hPTH vs. PTHrp.

Changes of systemic and regional hemodynamics during administration of hPTH (15 μg/kg/min) or PThrp (5 μg/kg/min) are shown in table 1. MAP and total peripheral resistance (TPR) significantly decreased and the HR of both groups increased significantly. Cardiac output increased significantly during the administration of hPTH, but PTHrp caused no significant change. Thus, the change of cardiac output with hPTH was larger than that with PTHrp and the change of TPR was also larger with hPTH (fig. 2).

The administration of hPTH (15 μg/kg/min) markedly increased blood flow in the heart, liver, and skin, but decreased it in the stomach, mesenterium, kidney, and spleen. The infusion of hPTH (15 μg/kg/min) decreased vascular resistance in the brain, heart, kidney, liver and skin, whereas it did not change in the stomach and mesenterium. In the PTHrp group, the changes of regional blood flow and vascular resistance were similar to those in the hPTH group, except that there was significant increase in skin blood flow (table 2; fig. 3).

Table 2. Effects of hPTH or PTHrp on regional hemodynamics in conscious rats

	hPTH (15 µg/kg/min), n = 7		PTHrp (5 µg/kg/min), n = 7	
	preinjection	postinjection	preinjection	postinjection
Regional blood flow, ml/min/g				
Brain	1.22 ± 0.08	1.37 ± 0.13	0.95 ± 0.15	0.85 ± 0.06
Heart	4.28 ± 0.58	9.78 ± 1.33**	3.49 ± 0.45	7.18 ± 0.29**
Liver	0.07 ± 0.02	0.42 ± 0.04*	0.08 ± 0.02	0.29 ± 0.05***
Spleen	2.36 ± 0.17	0.99 ± 0.14**	4.93 ± 0.66	0.61 ± 0.09**
Kidney	6.29 ± 0.29	5.40 ± 0.26*	4.93 ± 0.66	4.21 ± 0.25
Skin	0.17 ± 0.02	0.37 ± 0.07*	0.23 ± 0.04	0.27 ± 0.07
Stomach	1.39 ± 0.20	1.10 ± 0.15*	0.90 ± 0.11	0.64 ± 0.04*
Mesenterium	1.32 ± 0.12	0.93 ± 0.08**	0.77 ± 0.07	0.61 ± 0.06*
Muscle	0.26 ± 0.5	0.31 ± 0.06	0.16 ± 0.05	0.10 ± 0.01
Vascular resistance, mm Hg/ml/min/g				
Brain	89.6 ± 8.2	56.4 ± 5.7**	126.5 ± 17.1	91.3 ± 7.3**
Heart	27.5 ± 3.1	8.5 ± 1.3**	34.2 ± 4.9	10.6 ± 0.6**
Liver	2,393.4 ± 622.3	193.1 ± 31.1**	3,390.2 ± 1,638.7	335.4 ± 71.7**
Spleen	46.7 ± 4.4	84.8 ± 12.3*	74.1 ± 9.2	143.5 ± 21.1**
Kidney	17.0 ± 0.7	13.6 ± 0.6*	24.1 ± 3.3	18.2 ± 1.1*
Skin	704.9 ± 72.6	244.6 ± 44.3**	530.8 ± 66.5	395.7 ± 80.2*
Stomach	86.0 ± 10.5	75.2 ± 9.9	127.7 ± 14.3	120.9 ± 7.7
Mesenterium	84.3 ± 7.7	80.5 ± 5.9	143.3 ± 14.1	133.6 ± 16.8
Muscle	588.7 ± 162.4	338.5 ± 84.5*	1,120.5 ± 291.3	804.0 ± 77.2

Values are the means ± SEM. *p < 0.05, **p < 0.01, compared to preinjection values.

Discussion

Both hPTH and PTHrp reduced MAP in a dose-dependent manner in conscious rats, with the hypotensive effect of PTHrp being three times more potent than that of hPTH. This finding was in agreement with that of a previous study [11]. There was no significant difference in the reduction of MAP achieved by 15 μg/kg/min infusion of hPTH and 5 μg/kg/min of PTHrp. Cardiac output significantly increased in the hPTH group. This finding is in agreement with the report of Wang et al. [4]. In the PTHrp group cardiac output did not change despite the increase in HR and the reduction in afterload. A positive inotropic effect of PTHrp has been observed in the isolated perfused rat heart [11]. So, one possible explanation for the difference may be the reduction of preload by a decrease in venous return or diastolic filling time.

A vasodilatory effect of PTH on the vascular beds of the heart, kidney, liver and brain has been reported [4, 5]. However, in this study the changes of regional blood flow and vascular resistance suggested that a marked vasodilatory effect of hPTH only occurred in the vascular beds of the heart, liver, and skin. The lack of a significant increase in blood flow to the brain, kidney, stomach, and mesenterium appeared to be due to a reactive increase in pressor activity, such as a response of the sympathetic nervous and renin-angiotensin systems. The effects of PTHrp on regional hemodynamics were quite similar to those of hPTH, with both peptides causing a prominent vasodilator effect on the coronary and hepatic arteries.

Summary

PTHrp was discovered as a humoral hypercalcemic factor of malignancy and has been shown to bind the same receptor as PTH in rat bone cells, canine renal membranes, and rabbit renal microvessels. We investigated the global effect of human PTH(hPTH) and PTHrp on systemic and regional hemodynamics in conscious rats. The hypotensive response to PTHrp was more potent than that to hPTH. Although hPTH (15 μg/kg/min, i.v.) caused a significant increase in cardiac output, whereas PTHrp (5 μg/kg/min, i.v.) caused no change in cardiac output despite a similar hypotensive effect to hPTH, the effects of PTHrp and hPTH on regional hemodynamics were quite similar, and both peptides had a prominent vasodilatory effect on the coronary and hepatic arteries. Therefore, PTHrp appears to have an important role in blood pressure and regional hemodynamics as does hPTH.

References

1   Charbon GA: A rapid and selective vasodilator effect of parathyroid hormone. Eur J Pharmacol 1986;3:275–278.
2   Pang PKT, Tenner TE Jr, Lee JA, Yang M, Jansen HF: Hypotensive action of

parathyroid hormone preparations on rats and dogs. Proc Natl Acad Sci USA 1980;77:657–678.

3   Nickols GA, Merts MA, Cline WH: Endothelium-independent linkage of parathyroid hormone receptors of rat vascular tissue with increased adenosine 3′,5′-monophosphate and relaxation of vascular smooth muscle. Endocrinology 1986;119:349–356.

4   Wang H, Drruge ED, Yen Y, Blumenthal MR, Pang PKT: Effect of synthetic parathyroid hormone on hemodynamics and regional blood flows. Eur J Pharmacol 1984;97:209–215.

5   Nickols GA, Cline WH Jr: Vasodilation of canine vascular beds by parathyroid hormone. Pharmacologist 1983;25:724.

6   Class MF, Moore PL, Strickland ML, Pang PKT, Citak MS: Cardiovascular responses to parathyroid hormone. Am J Physiol 1985;249:E187–E194.

7   Katoh Y, Klein KL, Kaplan RA, Sanborn WG, Krokawa K: Parathyroid hormone has a positive inotropic action in the rat. Endocrinology 1981;109:2252–2254.

8   Suva LJ, Winslow GA, Wettenhall REH, Hammonds RG, Moseley JM, Diefenbach-Jagger H, Rodda CP, Kemp BE, Rodriguez H, Chen EY, Hudson PJ, Martin TJ, heart rate (HR) were monitored continuously through the femoral artery catheter. After the Wood WL: A parthyroid hormone-related protein implicated in malignant hypercalcemia: Cloning and expression. Science 1987;237:839–896.

9   Nisseson RA, Diep D, Strewler GJ: Synthetic peptides comprising the amino-terminal sequence of a parathyroid hormone like protein from human malignancies. J Biol Chem 1988;263:12866.

10 Musso MJ, Plante M, Judes C, Barthelmebs M, Helwig JJ: Renal vasodilation and microvessels adenylate cyclase stimulation by synthetic parathyroid hormone-like protein fragments. Eur J Pharmacol 1989;174:139–151.

11 Nickols GA, Nana AD, Nickols MA, Dipette DJ, Asimakis GK: Hypotention and cardiac stimulation due to the parathyroid hormone-related protein, humoral hypercalcemia of malignancy factor. Endocrinology 1989;125:834–841.

12 Ishise S, Pergram BL, Yamamoto J, Kitamura Y, Frohlich ED: Reference sample microsphere method: Cardiac output and blood flows in conscious rat. Am J Physiol 1980;239:H443–H449.

Hiroshi Kishimoto, MD, Second Department of Internal Medicine, Osaka City University Medical School, 1-5-7, Asahi-machi, Abeno-ku, Osaka 545 (Japan)

Morii H (ed): Calcium-Regulating Hormones. I. Role in Disease and Aging.
Contrib Nephrol. Basel, Karger, 1991, vol 90, pp 79–87

Effects of $1,25(OH)_2D_3$ on Cardiovascular Function

H. Jahn[a], *D. Schohn*[a], *A. Omichi*[a], *R. Miller*[b]

[a]Nephrology Department, Louis-Pasteur University, and [b]Merrell Dow Research Institute, Strasbourg, France

The presence of $1,25(OH)_2D_3$ receptors in the myocardium implies a regulatory role of calcitriol on the myocardial cells [1]. The receptor-mediated effects seem primarily related to de novo synthesis of proteins as has been described for steroid hormones. This means that calcitriol induces only delayed responses to the receptor stimulation. In heart cell cultures, Ca^{2+} uptake increases after 4 h when $1,25(OH)_2D_3$ is added in the media [2]. Eventual functional consequences would therefore occur after this delay.

Evidence has nevertheless been reported that other modes of action may exist. In vascular smooth muscle cell cultures, Ca^{2+} uptake increases within 15 min when $1,25(OH)_2D_3$ is added and seems therefore not to be related to de novo protein synthesis [3]. In rats, Ca^{2+} accumulation in the myocardium after vitamin D_3 administration is prevented by calcium-channel blockers which indicate the inhibition of other types of calcium influx [2].

The precise consequences of calcitriol myocardial regulation have not yet been elucidated. Long-term administration of vitamin D_3 induces modifications often related to extracellular Ca level modifications. Within therapeutic ranges $1,25(OH)_2D_3$ given to hemodialysis patients improves left ventricular function [4]. Amelioration of echocardiographic parameters in uremic patients after $1\alpha(OH)D_3$ have been reported [5]. Both these effects seem indirect and rather dependent on extracellular calcium and/or parathormone levels. Implications of direct effects on the myocardium, independent of calcium levels, appear in vitamin D_3 depletion experiments: vitamin D_3 deprivation induces cardiac hypertrophy due to interstitial edema, increment of collagen and decrease of myofibrillar area [6].

The cardiac effects of $1,25(OH)_2D_3$ need more investigations to get an overview of its role. In order to obtain further information, we studied the immediate effects with pharmacologic doses of $1,25(OH)_2D_3$ on cardiovascular function in dogs and on the Langendorff heart preparation.

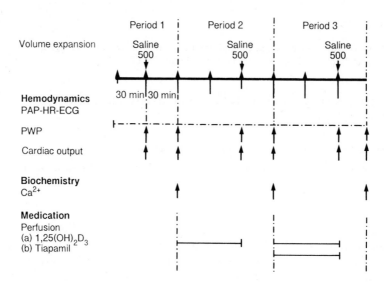

Fig. 1. Design of the study. Arrows indicate when the respective examinations are performed.

Methodology

The pharmacologic effects of $1,25(OH)_2D_3$ on cardiovascular function have been explored by the following investigations.

In vivo Studies

(1) *Modifications of cardiac function curves:* Five well-*trained* beagles with normal renal and cardiac function (24 ± 3 kg BW) were submitted, unanesthetized, to the experimental protocol designed in figure 1. Pulmonary artery catheterization was performed with a Swan-Ganz catheter (5F): pulmonary artery pressure (PAP), pulmonary wedge pressure (PWP), and cardiac output (CO) by thermodilution were measured. To obtain pressure and output modifications 0.9% saline was infused within 30 min in order to establish the function curves following Frank and Starling [7].

Period 1: Served to evaluate the basic cardiac function at the start of the investigations.

Period 2: A solution of $1,25(OH)_2D_3$ (total dose 20 µg) was infused and followed by saline infusion and measurement of hemodynamic parameters.

Period 3: Both $1,25(OH)_2D_3$ and Tiapamil were perfused and followed by saline infusion and measurement of hemodynamic parameters.

Control experiments comprised: (a) only saline infusions during the three periods; (b) replacement of $1,25(OH)_2D_3$ by the solvent and by Tiapamil; (c) administration of Ca instead of $1,25(OH)_2D_3$.

(2) *Modifications of left ventricular pressures, dp/dt and V_{max}:* The second series of experiments were studied in the same dogs anesthetized with Droleptan for each experimental protocol. Left ventricular catheterization was performed with a 5F Microtip Milar catheter.

Table 1. Variations of the hemodynamic parameters before and after saline expansion

	Period 1	Period 2 1,25(OH)$_2$D$_3$	Period 3 1,25(OH)$_2$D$_3$ +Tiapamil
ΔPWP, (mm Hg)	+7	+4	+1
ΔCl, liters/min/m^2	+1.3	+3.6	+0.8
ΔSI, ml/beat/m^2	+7.9	+22.7	−0.4
ΔHR, bpm	+10	+18	+15

The left ventricular pressure and dp/dt were recorded permanently (Multichannel Recorder Hellige with pressor integrator) [7] during 180 min. The first period of 60 min was utilized as a control period. During the second period of 60 min 1,25(OH)$_2$D$_3$ was perfused; during the third period of 60 min both 1,25(OH)$_2$D$_3$ and Tiapamil were perfused. Similar studies were performed with calcium and the ionophore Bay K 8644 in replacement of 1,25(OH)$_2$D$_3$. A dose-response curve was established with the following doses of 1,25(OH)$_2$D$_3$: 1, 5, 10, 15 and 20 µg (1,25(OH)$_2$D$_3$ for experimental usage and Tiapamil have been provided by Hoffmann-LaRoche Laboratories, Basel, and Bay K 8644 by Bayer Laboratories, Wuppertal).

(3) *An isoproterenol test was performed* to study the modifications of the sensitivity of cardiac beta-adrenergic receptors before and after 1,25(OH)$_2$D$_3$. It is performed with bolus injections of increasing doses of proterenol [8, 9].

(4) *Arteriolar reactivity* to infusion of angiotensin and norepinephrine before and during 1,25(OH)$_2$D$_3$ perfusion [9].

In vitro effects on the rabbit Langendorff heart preparation perfused with Tyrode's solution on the electrically stimulated taenia caeci preparation were investigated (this study was performed in the Merrel-Dow Laboratories).

Results

Hemodynamic Parameters and Cardiac Function Curves

The administration of 1,25(OH)$_2$D$_3$ induces several modifications: PWP does not change significantly but cardiac output increases by 22% (table 1). The cardiac function curves established from stroke index (SI) and PWP show a steeper curve due to an increment of 180% of stroke index in comparison to the stroke index before 1,25(OH)$_2$D$_3$ administration (fig. 2). This enhancement of the stroke index was blocked by the simultaneous Tiapamil administration. In the absence of changes of the afterload one may speculate that the results indicate a change in contractile state due to enhancement of myocardial performance. These modifications were observed whereas plasmatic calcium levels diminished (table 2). The variations of the hemodynamic parameters between period 1 and 2 are displayed in table 1.

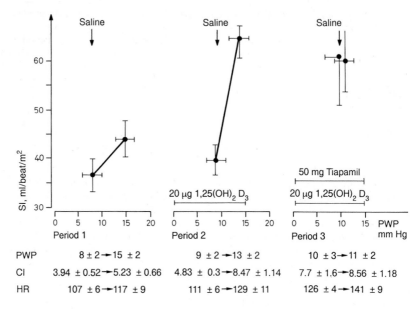

Fig. 2. Cardiac function curves in 5 dogs (mean ± SE).

Table 2. Ca, Mg and PO_4 modifications at the end of each period of the study

	Period 1 saline	Period 2 saline 20 μg $1,25(OH)_2D_3$	Period 3 saline 20 μg $1,25(OH)_2D_3$ 50 mg Tiapamil
Ca^{2+} mg/l	47.1 ± 5.1	43.6 ± 3.3	41.4 ± 3.1
Ca, mg/l	111 ± 9	93 ± 6	89 ± 4
Mg, mg/l	22.7 ± 3	17.7 ± 3	14.7 ± 2
PO_4, mg/l	40 ± 8	31 ± 7	26 ± 6
PTH, pg/ml	78 ± 27	52 ± 28	21 ± 2.9

Control studies have shown that similar effects can be obtained after calcium administration which nevertheless induced increments of plasma Ca levels to nearly 60% of its initial value.

The control experiments with saline alone, solvent of $1,25(OH)_2D_3$, do not show any modifications of stroke index period 2 and 3.

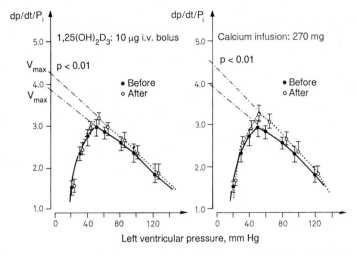

Calcium, mg/l: 101 ± 9 ➝ 93 ± 6   Calcium, mg/l: 100 ± 5 ➝ 153 ± 14

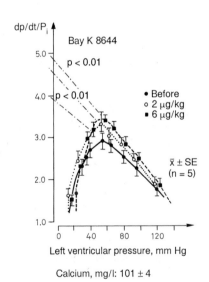

Fig. 3. dp/dt/P$_i$ modifications during each period of the study with extrapolation to V$_{max}$ before and after calcitriol, calcium and Bay K 8644.

Left Ventricular Function

After administration of 1,25(OH)$_2$D$_3$, the left ventricular pressure does not change significantly but dp/dt increases and is significantly different at the pic value (fig. 3). Similar effects are obtained after calcium or Bay K 8644 administration. The modifications of dp/dt are blocked by

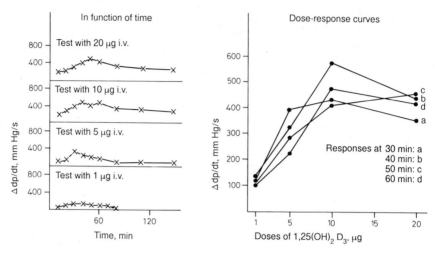

Fig. 4. Dose-response curves.

administration of Tiapamil. In these experimental conditions $1,25(OH)_2D_3$ showed increases of heart rate with arrhythmias, which were exacerbated by the simultaneous administration of Tiapamil. The variations of dp/dt to changing doses of $1,25(OH)_2D_3$ are dose dependent and approach linearity from 1 to 10 μg (fig. 4).

The Isoproterenol Test

$1,25(OH)_2D_3$ determines a shift to the left of the response curve to isoproterenol ($p < 0.05$). This indicates a slight increase of the heart sensitivity to this drug and implies an enhancement of the catecholamines effects on the heart (fig. 5).

The Arteriolar Reactivity to Norepinephrine and Angiotensin II

The response curves of both substances are shifted to the left indicating an increase of the pressure responses to endogenous vasopressure substances. These responses are different from those obtained after parathormone administration which is a vasorelaxant.

The in vitro Studies

The investigations performed on the Langendorff rabbit heart preparation did not show any effect of $1,25(OH)_2D_3$ at different doses (from 1 to

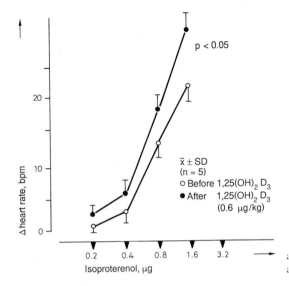

p < 0.05

$\bar{x} \pm SD$
(n = 5)
○ Before 1,25(OH)$_2$ D$_3$
● After 1,25(OH)$_2$ D$_3$
        (0.6 µg/kg)

Δ heart rate, bpm

20 —

10 —

0 —

0.2   0.4   0.8   1.6   3.2   ⟶

Isoproterenol, µg

Fig. 5. Heart rate modifications after isoproterenol, before and after calcitriol administration.

10 µg/l); whereas isoprenaline, Bay K 8644 and PTH elicit dose-dependent effects on contractility. Similar results have been obtained with the different substances when tested on the electrically stimulated taenia caeci.

Discussion

Administration of pharmacologic doses of 1,25(OH)$_2$D$_3$ enhances the contractile state of the myocardium in the dog. This is displayed by the investigation of the cardiac function curves, where we observed a majoration of the cardiac output by 22%. Saline expansion produces an increment of the stroke index of 22.7 ml per beat, which represents an increment of stroke index of 187% when compared to the period before the 1,25(OH)$_2$D$_3$ administration and the dp/dt recording displays a significant pressure increment in function of time. Both effects are related to the heart excitation contraction coupling and depend on the cellular movements of Ca^{2+}. As extracellular Ca^{2+} decreases in the conditions of our experimental protocol, the observed effects are not dependent on extracellular Ca^{2+} levels. Similar effects on dp/dt can be obtained by increment of extracellular Ca^{2+} and by the calcium ionophore Bay K 8644. Conversely, the calcium-channel blocker Tiapamil inhibits the majoration of the systolic

Table 3. Hemodynamic modifications before and during calcitriol administration

| | 1,25(OH)$_2$D$_3$ | |
	before	during
Arterial pressure, mg Hg		
Systolic	131 ± 10	124 ± 7
Diastolic	71 ± 9	68 ± 9
Heart rate, bpm	122 ± 16	140 ± 16*
Left ventricular pressure, mm Hg		
Diastolic	1 ± 2	1 ± 2
Telediastolic	5 ± 3	3 ± 2*
Systolic (P$_i$)	136 ± 11	127 ± 6
dp/dt$_{max}$, mm Hg/s	1,700 ± 240	2,240 ± 280*
dp/dt/P$_i$	12.5	17.6
V$_{max}$/s	38.6 ± 2.3	42.7 ± 1.9*

*$p < 0.05$.

index. The administration of 1,25(OH)$_2$D$_3$ therefore induces in vivo effects that seem related to cellular Ca^{2+} movements (table 3).

The effects on contractility are not found in the pharmacologic in vitro experiments: these effects therefore seem due to the modulation of other inotropic active substances or related to the presence of a substance which facilitates the effects of 1,25(OH)$_2$D$_3$.

In vivo, the effects of 1,25(OH)$_2$D$_3$ are observed within 30–60 min. They cannot depend on the described receptor-mediated effect which needs de novo protein synthesis. There must therefore exist a direct or indirect action on other mechanisms that promote the Ca^{2+} movements related to the excitation contraction coupling system. The effects of Ca channel blockers would point toward an activation of the Ca channels, or to a facilitation of the substances acting on Ca channels.

Some similarities with the inotropic action of PTH [10] on heart function appear, but up to now there are no arguments that 1,25(OH)$_2$D$_3$ has the deleterious effects on myocardial cells in uremic patients which have been elucidated by Massry and co-workers [11–13]. Heart calcinosis with a complex pattern of myocardial damage when observed have been described with overdoses of vitamin D$_3$ and severe hypercalcemia [14].

Summary

1,25(OH)$_2$D$_3$ enhances myocardial contractility with pharmacological doses from 1 to 20 μg within 30–60 min. dp/dt modifications induced by 1,25(OH)$_2$D$_3$ are dose dependent and approach linearity from 1 to 10 μg. The inotropic effects are not dependent on extracellular Ca^{2+} levels; but cellular Ca^{2+} influx or internal Ca^{2+} movement may be implied. The effect seems not related to the described 1,25(OH)$_2$D$_3$ receptors which implies protein de novo synthesis. A modulation on other receptor-mediated substances acting on Ca^{2+} movements seems probable.

References

1   Simpson RV: Evidence for a specific 1,25-dihydroxy vitamin D$_3$ receptor in rat heart. Circulation 1983;68:239.
2   Simpson RV, Weishaar RE: Involvement of 1,25-dihydroxyvitamin D$_3$ in regulating myocardial metabolism. Physiological and pathological actions. Cell Calcium 1988;9:285–292.
3   Tsutomu I, Hiroyuki K: 1,25-Dihydroxyvitamin D$_3$ stimulates $^{45}Ca^{24}$ uptake by cultured vascular smooth muscles derived from rat aorta. Biochem Biophys Res Commun 1988;152:1388–1394.
4   Coratelli P, Petrarulo F, Buongiorno E, Giannatasio M, Antonelli G, Amerio A: Improvement in left ventricular function during treatment of hemodialysis patients with 25-OH-D$_3$. Contr Nephrol. Basel, Karger, 1984, pp. 433–437.
5   McGonigle RJS, Fowler MB, Timmis AB, Weston MJ, Parsons V: Uremic cardiomyopathy: Potential role of vitamin D and parathyroid hormone. Nephron 1984;36:94–100.
6   Weishaar RE, Sang-Nam K, Dwight ES, Robert US: Involvement of vitamin D$_3$ with cardiovascular function. III. Effects on physical and morphological properties. Am J Physiol 1990;258:E134–E142.
7   Jahn H, Schmitt R, Schohn D, Olier P: Aspects of the myocardial function in chronic renal failure. Contr Nephrol. Basel, Karger, 1984, vol 41, pp. 240–250.
8   Cleaveland CR, Rangno RE, Shand DG: A standardized isoproterenol sensitivity test. The effects of sinus arhythmia, atropine and propranolol. Arch Intern Med 1972:130–147.
9   Schohn D, Weidmann P, Jahn H, Beretta-Piccoli C: Norepinephrine-related mechanism in hypertension accompanying renal failure. Kidney Int 1985;28:814–822.
10 Jahn H, Schohn D, Schmitt R: Plasma parathormone levels (PTH) and heart function in end-stage renal failure patients (Pts). Abstracts, IXth Int Congr Nephrology, 1984.
11 Bogin E, Massry SG, Harary I: Effect of parathyroid hormone on rat heart cells. J Clin Invest 1981;67:1215–1227.
12 Bogin E, Levi J, Harary I, Massry SG: Effects of parathyroid hormone on oxidative phosphorylation of heart mitochondria. Miner Electrolyte Metab 1982;7:151–156.
13 Baczinski R, Massry SG, Kohan R, Magott M, Saglikes Y, Brautbar N: Effect of parthyroid hormone on myocardial energy metabolism in the rat. Kidney Int 1985;27:718–725.
14 Wrzolkow AT, Zydowo M: Ultrastructural studies on the vitamin D-induced heart lesions in the rat. J Mol Cell Cardiol 1980;12:1117–1133.

Prof. Henry Jahn, Service de Nephrologie, Hospices Civils, 1, place de l'Hôpital, F–67091 Strasbourg (France)

Morii H (ed): Calcium-Regulating Hormones. I. Role in Disease and Aging.
Contrib Nephrol. Basel, Karger, 1991, vol 90, pp 88–93

Intracellular Ions in Salt-Sensitive Essential Hypertension: Possible Role of Calcium-Regulating Hormones

Lawrence M. Resnick[a], *Raj K. Gupta*[b], *Richard Z. Lewanczuk*[c],
Peter K.T. Pang[c], *John H. Laragh*[a]

[a]Cardiovascular Center New York Hospital-Cornell Medical Center New York;
[b]Department of Physiology and Biophysics, Albert Einstein College of Medicine,
Bronx, N.Y., USA; [c]Department of Physiology, University of Alberta School
of Medicine, Edmonton, Alta, Canada

Two fundamentally related questions about salt-sensitive forms of human hypertension remain problematic: (1) Why do some, but not all subjects exhibit an abnormal elevation of blood pressure in response to increases in dietary salt intake? (2) What are the mechanisms by which the dietary salt signal is transduced into a blood pressure effect? Research has addressed these questions at two levels. One at the cellular level, has focussed on defining the intracellular ionic responses to salt loading [1, 2]. The other has investigated the role of circulating factors such as endogenous digitalis-like substances [3] inappropriate renin secretion [4], and stimulation of circulating 1,25-dihydroxyvitamin D (1,25-D) levels [5].

We have attempted to link these approaches by combining NMR spectroscopic analysis of intracellular ionic responses to salt loading, with measurements of the recently described pressor substance of parathyroid origin, parathyroid hypertensive factor (PHF), in salt-sensitive and salt-insensitive essential hypertensive patients both in the free living state, and in response to dietary salt loading [6, 7]. Our results demonstrate salt-induced alterations of intracellular sodium, pH, free magnesium, and free calcium levels, the latter resulting from a shift of extracellular calcium intracellularly. The data further suggest that both 1,25-D and PHF may contribute to this constellation of ionic charges, and thus mediate salt-sensitive hypertension.

Table 1. Effects of salt loading on RBC intracellular ions

Δ (HS-LS)	Subjects		
	all (n = 17)	SS (n = 8)	SI (n = 9)
ΔDBP, %	7.6 ± 3*	14.6 ± 3**	1 ± 2
ΔCa$_i^f$, %	27.0 ± 11	44.0 ± 7*	12.2 ± 15.0
ΔMg$_i^f$, %	-14.2 ± 7*	-30.9 ± 9**	-1.5 ± 5
ΔpH$_i$	-0.03 ± 0.01	-0.05 ± 0.01*	-0.02 ± 0.01
ΔNa$_i$, %	25.5 ± 8*	35 ± 17*	13.0 ± 1.9

*p < 0.05, **p < 0.01 vs. SI.
ΔDBP = Change in diastolic blood pressure; Ca$_i^f$ = RBC cytosolic free calcium; Mg$_i^f$ = RBC intracellular free magnesium; pH$_i$ = RBC intracellular pH; Na$_i$ = RBC intracellular sodium.

Materials and Methods

The data reported here represent the results of studies performed over the period 1984–1989, the methods of which have been described elsewhere [5–7]. Briefly, unmedicated essential hypertensive subjects (n = 17) were enrolled in a randomized, cross-over study of high and low dietary salt intakes (200 < UNaV < 50 mEq/day) each of 1 month duration. On each diet, subjects were evaluated after an overnight fast with a 24-hour urine specimen collected the previous day to verify sodium intake. Blood was drawn in the seated position for analysis of plasma hypertensive factor (PHF), 1,25-D, serum ionized calcium (Ca-io), and for NMR spectroscopic analysis of intracellular free magnesium (Mg$_i$), cytosolic free calcium (Ca$_i$), intracellular pH (pH$_i$) and intracellular sodium (Na$_i$) levels. Salt sensitivity (SS) was defined as a blood pressure on a high dietary salt intake (UNaV > 200 mEq/d) more than 5% greater than on the low dietary salt intake (UNaV < 50 mEq/d).

Blood collected for PHF was prepared, stored, and assayed as described in detail elsewhere [8]. ^{31}P, ^{23}Na, and ^{19}F-NMR spectroscopic techniques were used to analyze erythrocyte intracellular levels of Mg$_i$, pH$_i$, Na$_i$, and Ca$_i$, respectively, as described in the literature [9–11]. Serum concentrations of 1,25-D were measured by standard radioreceptor assay methods [12], and serum ionized calcium was analyzed by a calcium-specific ion electrode apparatus (Corning), using blood drawn and processed anaerobically.

Results

Effects of Salt on Ca-io and Intracellular Ions

Intracellular ionic indices on high vs. low salt intakes in salt-sensitive (SS) vs. salt-insensitive (SI) hypertensive subjects are summarized in table 1. For the group as a whole, high salt (HS) vs. low salt (LS) diets were associated with a rise in RBC Na$_i$ (25.5 ± 8%, p < 0.001, HS vs. LS) and in cytosolic-free calcium (Ca$_i$) (27.0 ± 11%, p < 0.01, HS vs. LS).

Conversely, intracellular pH_i (-0.03 ± 0.01, $p < 0.05$ HS vs. LS) and free Mg_i ($-14.2 \pm 7\%$, $p < 0.05$, HS vs. LS) fell. Although Na_i rose in all subjects with salt loading, it rose to a much greater extent in SS ($n = 8$, $35 \pm 17\%$) vs. SI individuals ($n = 9$, $13.0 \pm 1.9\%$, $p < 0.05$ vs. SS). Furthermore, the salt-induced fall in pH_i and the reciprocal rise and fall, respectively, of Ca_i and Mg_i, were only present in the SS individuals, in whom blood pressure rose significantly with salt loading. Indeed, independently of the definition of salt sensitivity, the ability of salt to elevate the blood pressure was inversely related to the basal pH_i ($p < 0.001$) and to the salt-induced change in Mg_i ($p < 0.01$). The lower the basal pH_i and the greater the salt-induced fall in Mg_i, the more the salt-elevated pressure. At the same time cytosolic free calcium was rising, serum ionized calcium levels were falling. The %ΔDBP was also closely and inversely related to the fall in serum ionized calcium ($r = -0.72$, $p < 0.01$) [5].

Effects of Salt on 1,25-Dihydroxyvitamin D and Parathyroid Hypertensive Factor

The ability of salt to elevate blood pressure was closely related ($r = 0.82$, $p < 0.05$) to its ability to stimulate circulating levels of 1,25-D [5]. Levels of PHF also differed in SS compared with SI subjects. SS individuals in the free living state on an unrestricted diet exhibited higher PHF levels than did SI subjects ($+362 \pm 40\%$, $p < 0.001$, SS vs. SI), despite similar average urinary sodium excretion levels. Furthermore, for a given subject studied on both dietary salt intakes, PHF rose significantly on the high vs. low salt diets ($325 \pm 31\%$, $p < 0.001$, HS vs. LS). Lastly, the ability of salt to elevate pressure on high vs. low dietary salt diet was closely related to the basal level of PHF – the higher the basal PHF, the more the pressure rose with salt loading ($p < 0.001$).

Discussion

The central findings of these studies are that extremes of dietary salt intake are associated with: (1) steady state alterations of erythrocyte intracellular Na_i, pH_i, Ca_i, and Mg_i levels. Specifically, intracellular Na_i and Ca_i rose, while pH_i and Mg_i fell in response to salt loading. (2) This altered cellular ion handling is closely related to salt-induced elevation of blood pressure, i.e. the greatest pressure reponses to salt were found in those patients exhibiting the greatest alterations of intracellular ion levels. This suggests that the changes observed here in the red blood cell may be physiologically relevant to and reflective of changes in other tissues such as vascular smooth muscle. (3) Concomitantly with the above ionic changes,

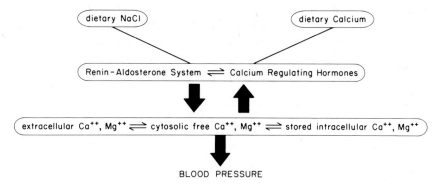

Fig. 1. Overall scheme in which environmental dietary mineral inputs are transduced at the cellular level by the renin-aldosterone system and calcium-regulating hormones. The resultant changes in intracellular vs. extracellular ion content mediate the blood pressure consequences of altered dietary sodium and/or calcium intake.

a suppression of serum ionized calcium is noted, inversely related to the pressor response to salt, suggesting that salt induces cellular calcium uptake from the extracellular space in salt-sensitive subjects. Similar results have recently been reported by Japanese workers [1]. (4) Perhaps explaining these linked extracellular and intracellular ionic changes, are parallel elevations of 1,25-D and of PHF induced by salt loading. These calcium-related hormones, each capable of stimulating calcium uptake into vascular smooth muscle tissue [13, 14], provide at least one mechanism by which the extracellular dietary sodium signal is translated at the cellular level to evoke a pressor response. Elevated levels of these hormones among essential hypertensive subjects may thus be a marker for the salt-sensitive stage, and may also provide a possible mechanism whereby dietary salt loading elevates blood pressure in salt-sensitive subjects.

This hypothesis is supported by our previous work in which calcium supplementation offset the pressor effects of salt proportionately to the degree it reversed salt-induced elevations of 1,25-D [15]. The ability of calcium supplementation to suppress circulating PHF levels has also recently been reported in the SHR model [16]. Altogether, as depicted in figure 1, these data suggest that the different blood pressure responses to salt loading observed among different hypertensive subjects may derive from and/or be mediated by different metabolic set points of critical calcium regulatory hormones such as 1,25-D and PHF. If elevated, these hormones would promote cellular calcium uptake from the extracellular space, accounting for the fall in serum ionized calcium and the rise in cytosolic free calcium, and, in turn, the rise in blood pressure observed among SS but not SI individuals. Further research needs to study the direct actions of these hormones on cellular ion handling, and to establish whether alterations of these hormones are necessary and/or sufficient conditions for the occurrence of the salt sensitivity in man.

Summary

To study the blood pressure, ionic, and hormonal responses to chronic dietary salt loading, we measured RBC pH_i, Mg_i, Na_i, and Ca_i, and compared these with serum levels of Ca-io, 1,25-D, and free levels of PHF in salt-sensitive and salt-insensitive essential hypertension subjects on low and high (200 mEq/day $<$ UNaV $<$ 50 mEq/day) dietary salt intakes.

As a group, salt loading significantly increased diastolic blood pressure ($7.6 \pm /3\%$), Na_i ($25.5 \pm /8\%$) and Ca_i ($27.0 \pm /11\%$), while Mg_i ($-14.2 \pm 7\%$) and pH_i ($-0.03 \pm /0.01$ pH units) fell. Significant changes in pH_i, Ca_i, and Mg_i occurred only in SS individuals, who also exhibited elevated levels of 1,25-D and of PHF in association with suppression of Ca-io. Altogether, the pressor response to salt was inversely related to the basal pHi ($p < 0.001$), and to the degree of salt-induced suppression of Ca-io ($p < 0.001$) and Mg_i ($p < 0.001$), while being directly related to the basal PHF ($p < 0.001$) and to the salt-induced stimulation of 1,25-D ($p < 0.005$).

We hypothesize that elevated levels of 1,25-D and PHF coordinately shift intracellular cation levels and stimulate cellular calcium uptake from the extracellular space. As such, these calcium-regulating hormones may be responsible, at least in part, for salt-sensitive hypertension.

References

1   Oshima T, Matsuura H, Matsumoto K, Kido K, Kajiyama G: Role of cellular calcium in salt sensitivity of patients with essential hypertension. Hypertension 1988;11:704–707.

2   Nicholson JP, Resnick LM, Cigarroa J, Marion D, Vaughan ED Jr, Laragh JH: The pressor effect of sodium-volume expansion is calcium-mediated. Clin Res 1989;37:397A.

3   MacGregor GA, deWardener HE: Is a circulating sodium transport inhibitor involved in the pathogenesis of essential hypertension? Clin Exp Hypertens 1981;3:815–830.

4   Sealey JE, Blumenfeld JD, Bell GM, Pecker MS, Sommers SC, Laragh JH: On the renal basis for essential hypertension: nephron heterogeneity with discordant renin secretion and sodium excretion causing hypertensive vasoconstriction-volume relationship. J Hypertens 1988;6:763–779.

5   Resnick LM, Nicholson JP, Laragh JH: Alterations in calcium metabolism mediate dietary salt sensitivity in essential hypertension. Trans Assoc Am Phys 1985;98:313–321.

6   Resnick LM, Gupta RK, Laragh JH: Role of intracellular cations in dietary salt sensitivity. Clin Res 1990;38:476A.

7   Resnick LM, Lewarczuk RZ, Laragh JH, Pang PKT: Plasma hypertensive factor in essential hypertension. Clin Res 1990;38:477A.

8 Lewanczuk RZ, Resnick LM, Blumenfeld JD, Laragh JH, Pang PKT: A new circulating hypertensive factor in the plasma of essential hypertensive subjects. J Hypertens 1990;8:105–108.

9 Resnick LM, Gupta PK, Laragh JH: Intracellular free magnesium in erythrocytes of essential hypertension: Relation to blood pressure and serum divalent cations. Proc Natl Acad Sci USA 1984;81:6511–6515.

10 Resnick LM, Gupta RK, Sosa RE, Corbett ML, Laragh JH: Intracellular pH in human and experimental hypertension. Proc Natl Acad Sci USA 19087;84:7663–7667.

11 Gupta RK, Gupta P, Moore RD: NMR studies of intracellular metal ions in intact cells and tissues. Ann Rev Biophys Bioeng 1984;13:221–246.

12 Resnick LM, Müller FB, Laragh JH: Calcium-regulating hormones in essential hypertension. Ann Intern Med 1986;105:649–654.

13 Merke J, Hofman W, Goldenschmidt I, Ritz E: Demonstration of $1,25(OH)_2$ vitamin D_3 receptors and actions in vascular smooth muscle cells in vitro. Calcif Tissue Int 1987;41:112–114.

14 Lewanzcuk RZ, Resnick LM, Blumenfeld JD, Laragh JH, Pang PKT: A new circulating hypertensive factor in the plasma of essential hypertensive subjects. J Hypertens 1990;8:105–108.

15 Resnick LM, DiFabrio B, Marion RM, James GD, Laragh JH: Dietary calcium modifies the pressor effects of dietary salt intake in essential hypertension. J Hypertens 1986;4(suppl 6):S679–S681.

16 Lewanzcuk RZ, Chen A, Pang PKT: The effects of dietary calcium on blood pressure in spontaneous hypertensive rats may be mediated by parathyroid hypertensive factor. Am J Hypertens 1990;3:349–353.

Lawrence W. Resnick, MD, Cardiovascular Center, The New York Hospital – Cornell Medical Center, 520 East 70th Street, room 420, New York, NY 10021 (USA)

Morii H (ed): Calcium-Regulating Hormones. I. Role in Disease and Aging.
Contrib Nephrol. Basel, Karger, 1991, vol 90, pp 94–98

Exaggerated Natri-Calci-Uresis and Increased Circulating Levels of Parathyroid Hormone and 1,25-Dihydroxyvitamin D in Patients with Senile Hypertension

S. Morimoto, M. Imaoka, S. Kitano, S. Imanaka, K. Fukuo, Y. Miyashita, E. Koh, T. Ogihara

Department of Geriatric Medicine, Osaka University Medical School, Osaka, Japan

Recent observations revealed aberration of calcium (Ca) metabolism in patients with essential hypertension: Resnick et al. [1] reported that there was a clear negative correlation between the plasma renin activity (PRA) and circulating level of 1,25-dihydroxyvitamin D [$1,25(OH)_2D$] in these patients. On the other hand, some of the patients with essential hypertenison have been known to show exaggerated natriuresis in response to saline infusion [2], although renal handling of other electrolytes has not been well elucidated. The kidney is known to play a central role in maintaining homeostasis of Ca and inorganic phosphate (P_i) in the renal tubules; major parts of reabsorption of Ca and P_i in the renal tubules were known to closely correlate with the reabsorption of sodium (Na) [3, 4]. In the present study, circulating levels of Ca-regulating hormones, including parathyroid hormone (PTH), calcitonin (CT), 25-hydroxyvitamin D (25-OHD), $1,25(OH)_2D$ and 24,25-dihydroxyvitamin D [$24,25(OH)_2D$], were evaluated in normotensive and hypertensive senile females, under comparison of renal handling of electrolytes in response to physiological saline infusion in these subjects.

Subjects and Methods

The study group included 44 normotensive (79.1 ± 4.1 years) and 27 hypertensive (mean ± SD age 79.8 ± 9.2 years) elderly females, the latter had mild-to-moderate essential hypertension with blood pressure readings above 150/90 mm Hg with stage I or II by the classification of the World Health Organization. All the subjects had an ordinary hospital diet containing about 120 mEq/day of Na, 600 mg/day of Ca, and 1,000 mg/day of Pi for at least 1 month. All the subjects were explained the nature of the study, and gave their written consent. None of them suffered from cardiovascular diseases, renal diseases, diabetes mellitus or other severe illness except for hypertension, nor was medicine administered which might influence Na, Ca or P_i metabolism. After an overnight fast, the subjects, in a supine position, were infused intravenously with 20 ml/kg of physiological saline from 9:00 a.m. to 11:00 a.m. Two 2-hour urine specimens were collected from 7:00 a.m. to 9:00 a.m. (U_1), and from 9:00 a.m. to 11:00 a.m. (U_2). Blood samples were obtained by venipuncture at 7:00 a.m. (B_1), 9:00 a.m. (B_2) and 11:00 a.m. (B_3).

The serum and urinary levels of Na, potassium (K), chloride (Cl), Ca, P_i and creatinine (Cr), and the serum level of albumin were measured with a multichannel autoanalyser. Endogenous creatinine clearance (C_{Cr}), and fractional excretion (FE) of each electrolyte were calculated by standard formulas [5] using the values of the mean values of the electrolytes in B1, B2 and that in U1 for the basal values and those in B2, B3 and in U2 for the saline infusion-responsive values. PRA and plasma aldosterone concentrations were measured by previously reported methods [6, 7]. Plasma levels of intact molecules of PTH and CT were measured with commercially available radioimmunoassay (RIA) kits [8, 9]. The serum levels of 250HD and 1,25(OH)$_2$D were measured by previously reported methods [10]. The serum level of 24,25(OH)$_2$D was determined by a competitive protein-binding assay [11]. Results were analyzed by two-way analysis of variance by the Bonferroni method.

Results

The hypertensive senile subjects showed statistically significant ($p < 0.01$) increases in the systolic (160 ± 11 vs. 119 ± 12 mm Hg), diastolic (91 ± 13 vs. 70 ± 5 mm Hg) and mean blood pressure (114 ± 11 vs. 86 ± 5 mm Hg), respectively. There was no statistically significant change in the blood pressure before and during the saline infusion in the two groups.

Compared to the normotensive group, the hypertensive group showed statistically significant decreases in the basal values of serum Na (141 ± 2 vs. 143 ± 3 mEq/l, $p < 0.05$) and Ca (4.6 ± 0.3 vs. 4.9 ± 0.4 mEq/l, $p < 0.01$) and of PRA (0.67 ± 0.81 vs. 1.23 ± 1.26 ng/ml/min, $p < 0.01$). However, the mean serum levels of albumin were similar between the two groups (3.66 ± 0.33 vs. 3.58 ± 0.27 g/dl). Figure 1 compares the circulating Ca-regulating hormones between the normotensive and hypertensive senile subjects. The group of hypertensive senile patients showed statistically significantly ($p < 0.05$) increased circulating levels of PTH and 1,25(OH)$_2$D and statistically significantly decreased levels of 24,25(OH)$_2$D. There was no statistically significant difference in the basal urine volume or urinary excretions of electrolytes in U1 between the two groups. In response to the saline infusion, the urine volume and urinary excretions of electrolytes including Na, K, Cl, Ca and P_i were increased significantly ($p < 0.05$) in the two groups compared to the respective basal values. Compared to the normotensive group, the hypertensive group showed statistically significant increases in the urinary excretions in U_2 of Na (275 ± 140 vs.

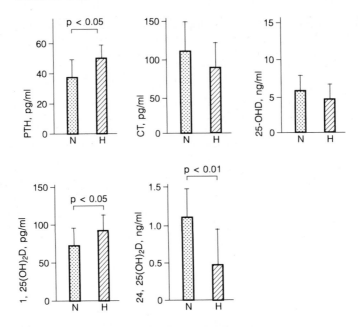

Fig. 1. Comparisons of calcium-regulating hormones between the normotensive (N) and hypertensive (H) senile female subjects.

$156 \pm 108 \, \mu\text{Eq/min}$, $p < 0.01$), Ca (1.26 ± 0.85 vs. $0.74 \pm 0.61 \, \mu\text{Eq/min}$, $p < 0.05$) and P_i (4.73 ± 2.91 vs. $2.95 \pm 1.86 \, \mu\text{g/min}$, $p < 0.05$).

Figure 1 compares the endogenous creatinine clearance (C_{Cr}) and fractional excretion (FE) of electrolytes in U_1 and in U_2 between the two groups. Compared to the normotensive group, the hypertensive group showed statistically significant increases in the FE values of Na and Ca in U1 and in U_2, and in those of P_i in U2.

Discussion

In the present study, senile subjects with essential hypertension were revealed to show significant decreases in basal serum levels of Na and Ca, and significantly increased levels of circulating levels of PTH and $1,25(\text{OH})_2\text{D}$ compared to age-matched senile female subjects (fig. 1). The increase in the circulating levels of these calcitropic hormones may be increased secondly due to the decrease in the serum level of Ca. These hypertensive patients also showed exaggerated natriuresis associated with excessive excretion of Ca and P_i in response to the saline infusion, probably

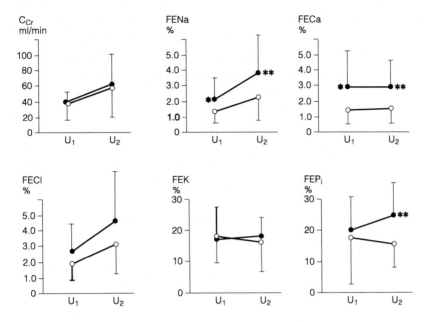

Fig. 2. Comparisons of the C_{Cr} and FENa, FECl, FEK, FECa and FEP_i of the 2 h before and during the saline infusion between the normotensive ($\bigcirc$) and hypertensive ($\bullet$) senile female subjects. * p < 0.05; ** p < 0.01.

through the above-mentioned Na-dependent Ca and P_i handling of the kidney. The excess excretion of Na, Ca and P_i during the saline infusion in the patients with hypertension was not caused by a particular increase in the glomerular filtration rate, but may be caused by decreased reabsorption of these electrolytes by the renal tubules (fig. 2). Excess loss of the Ca associated with the defect of Na reabsorption may be the cause of the significant decrease in the mean serum level of Ca and the increase in the circulating levels of PTH and $1,25(OH)_2D$ in these patients. Since exaggerated natriuresis has been observed in patients with essential hypertension with low PRA [2], and since senile subjects tend to show decreased level of PRA [12], these features may explain the aberration of Ca metabolism in senile hypertensive patients.

Summary

Renal handling of Na and Ca in response to physiological saline infusion (20 ml/kg i.v. for 2 h) was compared between 27 hypertensive (mean ±SD age 79.8 ± 9.2 years) and 44 normotensive (79.1 ± 4.1 years) senile females. Compared to the normotensive group, the hypertensive group showed statistically significant decreases in the basal values of serum Ca and PRA, and significant increases in basal circulating levels of parathyroid hormone and $1,25(OH)_2D$ and in urinary excretions of Na, Ca and P_i in the 2-hour urine specimens during the saline infusion. These results suggest that the excessive excretions of Ca and P_i associated with exaggerated natriuresis may participate in aberration of Ca metabolism in low-renin hypertensive seniles.

References

1   Resnick LM, Muller FB, Laragh JH: Calcium regulating hormones in essential hypertension. Ann Intern Med 1986;105:649–654.

2   Krakoff LR, Goodwin FJ, Baer L, et al: The role of renin in the exaggerated natriuresis of hypertension. Circulation 1970;17:335–345.

3   Duarte CG, Watson JF: Calcium reabsorption in proximal tubule of the dog nephron. Am J Physiol 1967;212:1355–1360.

4   Dennis VW, Woodhall PB, Robinson RR: Characteristics of phosphate transport in isolated proximal tubule. Am J Physiol 1976;231:979–984.

5   Eliison DH, Shneidman R, Morris C, McCarron DA.: Effects of calcium infusion on blood pressure in hypertensive and normotensive humans. Hypertension 1986;8:497–505.

6   Ogihara T, Iinura K, Nishi K, Arakawa Y, Takagi A, Kurata K, Miyai K, Kumahara Y: A non-chromatographic bib-extraction radioimmunoassay for serum aldosterone. J Clin Endocrinol Metab 1977;45:726–731.

7   Ogihara T, Shima J, Hara H, Tabuchi Y, Hashizume K, Nagano M, Katahira K, Kangawa K, Matsuo H, Kumahara Y: Significant increase in plasma immunoreactive atrial natriuretic polypeptide concentration during head-out water immersion. Life Sci 1986;38:2413–2418.

8   Nussbaum SR, Zahradnik RJ, Labigne JR, Brennan GL, Nozawa-Ung K, Kim LY, Keutmann HT, Wang C, Potts JT Jr, Segre GV: Highly sensitive tow-site immunoradiometric assay of parathyrin, and its clinical utility in evaluating patients with hypercalcemia. Clin Chem 1987;33:1364–1367.

9   Morimoto S, Tsuji M, Okada Y, Onishi T, Kumahara Y: The effect of oestrogens on human calcitonin secretion after calcium infusion in elderly female subjects. Clin Endocrinol 1980;13:135–143.

10 Lee S, Morimoto S, Onishi T, Tsuji Y, Okada Y, Seino M, Ishika M, Yamaoka K, Takai S, Miyauchi A, Kumahara Y: Normal serum 1,25-dihydroxyvitamin D in patients with medullary carcinoma of the thyroid. J Clin Endocrinol Metab 1982;55:361–363.

11 Seino Y, Tanaka H, Yamaoka K, Ishida M, Yabuuchi H: Circulating 1,25-dihydroxyvitamin D levels after a single dose of 1,25-dihydroxyvitamin D_3 or 1α-hydroxyvitamin D_3 in normal men. Bone Mineral 1987;2:479–485.

12 Weidmann P, De Myttenaere-Bursztein S, Maxwell MH, et al: Effect of aging on plasma renin aldostrone in normal man. Kidney Int 1975;8:325–333.

Shigeto Morimoto, MD, Department of Geriatric Medicine, Osaka University Medical School, Fukushima-ku, Osaka 553 (Japan)

Role of Endothelin

Morii H (ed): Calcium-Regulating Hormones. I. Role in Disease and Aging.
Contrib Nephrol. Basel, Karger, 1991, vol 90, pp 99–104

The Mechanisms of Endothelin Action in Vascular Smooth Muscle Cells

Yoh Takuwa, Tomoh Masaki, Kamejiro Yamashita

Department of Internal Medicine, Institute of Clinical Medicine and Department
of Pharmacology, Institute of Basic Medicine Sciences, University of Tsukuba,
Ibaraki, Japan

Endothelin-l (ET) is one of the most potent vasoconstrictors known so
far [1]. Previous studies demonstrated that ET stimulates inositol lipid
hydrolysis and Ca^{2+} channel gating in vascular smooth muscle and other
types of cells [1–6]. In recent years it has been shown that in most cases
receptor stimulation by Ca^{2+}-mobilizing agonists is coupled to phospholipase C activation via guanine nucleotide-binding regulatory proteins (G
proteins) [7]. At present, the nature of ET receptor(s) and the mechanism
by which the receptor activation is coupled to phospholipase C stimulation
are not known.

In the present study, we tried to characterize the receptor and
the signal transduction mechanisms of ET using cultured rat aortic vascular smooth muscle cells (A-10 cells). We concluded from the results
that ET binds to and activates a single class of high affinity receptors
which are coupled to phospholipase C via a pertussis toxin-insensitive G
protein.

Methods

A-10 cells were grown as described elsewhere [8]. The ^{125}I-ET binding experiments were
performed as described previously [6, 8, 9]. Affinity-crossing linking of ^{125}I-ET to intact cells
was carried out by employing a cross-linker, ethylene glycolbis (succinimidyl succinate) [8].

The intracellular free Ca^{2+} concentration ($[Ca^{2+}]_i$) was measured by employing a
fluorescent Ca^{2+} indicator, fura-2 [8, 9]. The production of inositol phosphates in
[^3H]inositol-prelabeled cells and in membranes prepared from [^3H]inositol-prelabeled cells
were measured as described previously [6, 8, 9]. The mass of cellular 1,2-diacylglycerol was
measured with an enzymatic method employing a diacylglycerol kinase [6, 8, 9].

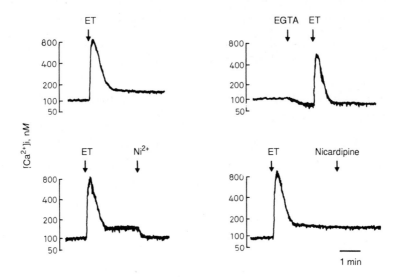

Fig. 1. Effect of ET on the intracellular free Ca^{2+} concentration. Test substances ($10^{-7} M$ ET, $2 \times 10^{-3} M$ EGTA, $1 \times 10^{-3} M$ Ni^{2+} and $1 \times 10^{-6} M$ nicardipine) were applied at arrows and were present throughout each recording.

Results

Intact A-10 cells shows specific binding of ^{125}I-ET which increases in a concentration-dependent manner. Scatchard analysis of the data reveals a single class of high affinity binding sites with a kD of $3 \times 10^{-10} M$ and a B_{max} of 67,000 binding sites per cell. Affinity cross-binding experiments demonstrate the presence of a single major band with an M_r of 65,000–75,000 on SDS-PAGE.

Addition of ET ($10^{-7} M$) induces a prompt rise in $[Ca^{2+}]_i$ which peaks within 15 s of ET addition (fig. 1). The $[Ca^{2+}]_i$ then falls to a lower plateau level. The peak $[Ca^{2+}]_i$ increase becomes greater with increasing doses of ET and reaches a maximal level at $3 \times 10^{-8} M$ ET. When extracellular Ca^{2+} is lowered by adding 2 mM EGTA (free Ca^{2+} 200 nM), the basal $[Ca^{2+}]_i$ declines to a new steady state level. Under such a condition, ET still induces the initial Ca^{2+} transient, but not the second plateau of the $[Ca^{2+}]$ response. In the presence of 1.25 mM $CaCl_2$, addition of 1 mM Ni^{2+}, but not $10^{-6} M$ nicardipine, promptly abolishes the second plateau.

As shown in figure 2, ET ($10^{-7} M$) induces marked increases in all three inositol phosphates in [^3H]inositol-prelabeled A-10 cells. Inositol

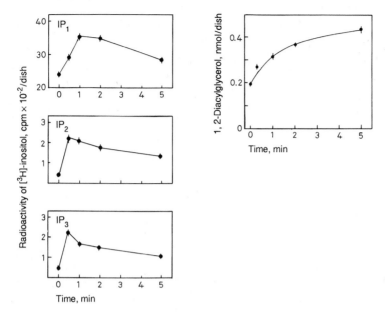

Fig. 2. Time-dependent changes of levels of inositol phosphates and 1,2-diacylglycerol in cells stimulated with 10^{-7} *M* ET. The data represent the mean $\pm$ SD of three determinations.

bis- and trisphosphates increase promptly, peak at 30 s and then gradually decline for the next 4 min. Inositol monophosphate rises less rapidly and peaks at 1 min. ET also induces increases in cellular 1,2-diacylglycerol (DAG) content. Within 20 s of ET (10^{-7} *M*) addition, cellular DAG content increases significantly and continues to rise over 5 min.

We next examined the effect of a nonhydrolyzable GTP analogue, guanosine 5′-0-(thiotriphosphate) (GTPrS) on ^{125}I-ET binding to A-10 membranes and ET-induced inositol phosphate production in A-10 cell membranes prepared from [^{3}H]inositol-prelabeled cells. GTPrS inhibits the specific binding of ^{125}I-ET to A-10 membranes in a dose-dependent manner. At 100 µM, GTPrS inhibits the ^{125}I-ET binding by 70%. In the absence of GTPrS, 10^{-7} *M* of ET alone does not significantly stimulate inositol phosphate production (fig. 3). GTPrS (10^{-6} *M*) by itself has a small stimulating effect on inositol phosphate production. However, if ET is applied together with GTPrS, marked increases in inositol phosphate production are induced. GTPrS enhances ET-induced production of inositol phosphate in A-10 membranes in a dose-dependent manner.

The exposure of intact A-10 cells to 10 ng/ml of pertussis toxin for 24 h does not significantly affect the subsequent ET-induced inositol

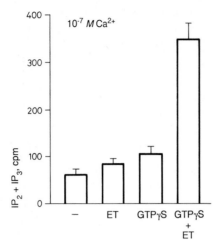

Fig. 3. Dependence of ET-stimulated inositol phosphate production on GTPrS. [^{3}H]inositol-prelabeled plasma membranes were incubated with 10^{-7} *M* ET in the presence or absence of 10^{-6} *M* GTPrS for 1 min at 37 °C. Free Ca^{2+} concentration in the assay buffer was 10^{-7} *M*. The data represent the mean ± SD of three determinations.

phosphate production. In cells pretreated with pertussis toxin the peak $[Ca^{2+}]_i$ increment induced by 10^{-7} *M* ET is smaller by only 20% than in nontreated cells, and the second plateau of the $[Ca^{2+}]_i$ response is comparable between pertussis toxin-treated and nontreated cells.

Discussion

The results in the present study demonstrate that ET activates phospholipase C to produce inositol trisphosphate and DAG by a G protein-dependent mechanism in cultured vascular smooth muscle cells. A-10 cells possess a single class of high-affinity receptor for ET. The binding of ET to the receptors on A-10 cell membranes is inhibited by GTPrS in a dose-dependent manner, suggesting that the ET receptor interacts with G proteins [10]. Moreover, ET stimulates polyphosphoinositide hydrolysis in A-10 cell membranes in a manner strictly dependent on GTPrS (fig. 3). These results strongly suggest that the ET receptor is coupled to phospholipase C by a G protein in A-10 cells. Pretreatment of A-10 cells with pertussis toxin fails to inhibit ET-induced inositol phosphate production. These findings suggest that a G protein distinct from pertussis toxin-sensitive G protein, Gi or Go, couples the ET receptor to phospholipase C in A-10 cells.

The measurements of $[Ca^{2+}]_i$ with fura-2 demonstrate a biphasic pattern of ET-induced Ca^{2+} mobilization: an initial transient increase in $[Ca^{2+}]_i$ due to intracellular mobilization and a second plateau phase. The second plateau phase of the $[Ca^{2+}]_i$ response to ET is dependent on the presence of extracellular Ca^{2+}, and is blocked by an inorganic Ca^{2+} channel blocker, Ni^{2+}, but not a dihydropyridien Ca^{2+} channel antagonist, nicardipine, suggesting that the second plateau phase is mainly due to Ca^{2+} influx through a Ca^{2+} channel which is distinct from the dihydropyridine-sensitive (L-type), voltage-dependent Ca^{2+} channel. The failure of pertusis toxin pretreatment to inhibit the second plateau of the $[Ca^{2+}]_i$ response to ET in A-10 cells suggests that activation of the Ca^{2+} channel by ET is not mediated by the pertussis toxin-sensitive G proteins.

In conclusion, the present results indicate that the receptor activation by ET is coupled to phosphoinositide hydrolysis through a pertussis toxin-insensitive G protein and gating of dihydropyridine-insensitive Ca^{2+} channel. It is suggested that these transmembrane signaling pathways play an important role in mechanisms for ET-induced vasoconstriction [11].

Summary

The mechanisms of actions were investigated in cultured rat aortic vascular smooth muscle A-10 cells. The A-10 cells have a single class of high affinity binding sites for ET with an apparent M_r of 65,000–75,000 on SDS-PAGE. Stimulation of cells with ET induces mobilization of Ca^{2+} from both intra- and extracellular pools to produce a biphasic increase in cytoplasmic free Ca^{2+} concentration. A dihydropyridine Ca^{2+} channel antagonist does not inhibit the second plateau phase of the $[Ca^{2+}]_i$ increase which is dependent on extracellular Ca^{2+}. ET stimulates phospholipase C to produce inositol trisphosphate and 1,2-diacylglycerol vai a pertussis toxin-insensitive G protein. These results indicate that the receptor activation by ET is coupled to phospholipase C activation and Ca^{2+} channel gating in vascular smooth muscle cells.

References

1 Yanagisawa M, Kurihara H, Kimura, S, Tomobe Y, Kobayashi M, Yazaki Y, Goto K, Masaki T: A novel potent vasoconstrictor peptide produced by vascular endothelial cells. Nature 1988;332:411–415.

2 Kasuya Y, Ishikawa T, Yanagisawa M, Kimura S, Goto K, Masaki T: Mechanism of contraction to endothelin in the isolated porcine coronary artery. Am J Physiol 1989;257:H1828–H1835.

3 Goto K, Kasuya Y, Matsuki N, Takuwa Y, Kurihara H, Ishikawa S, Kimura M, Yanagisawa M, Masaki T: Endothelin activates the dihydropyridine-sensitive, voltage-dependent Ca^{2+} channel in vascular smooth muscle. Proc Natl Acad Sci USA 1989;86:3915–3918.

4 Resink TJ, Scott-Burden T, Bühler FR: Endothelin stimulates phospholipase C in cultured vascular smooth muscle cells. Biochem Biophys Res Commun 1988;157:1360–1368.

5 Kasuya Y, Takuwa Y, Yanagisawa M, Kimura S, Goto K, Masaki T: Endothelin-1 induces vasoconstriction through two functionally distinct pathways in porcine coronary artery: Contribution of phosphoinositide turnover. Biochem Biophys Res Commun 1989;161:1049–1055.

6 Takuwa N, Takuwa Y, Yanagisawa M, Yamashita K, Masaki T: A novel vasoactive peptide endothelin stimulates mitogenesis through inositol lipid turnover in Swiss 3T3 fibroblasts. J Biol Chem 1989;264:7856–7861.

7 Martin TFJ, Bajjalieh SM, Lucas DO, Kowalchyk JA: Thyrotropin-releasing hormone stimulation of polyphosphoinositide hydrolysis in GH3 cell membranes in GTP dependent but insensitive to cholera or pertussis toxin. J Biol Chem 1986;261:10041–10049.

8 Takuwa Y, Kasuya T, Takuwa N, Kudo M, Yanagisawa M, Goto K, Masaki T, Yamashita K: Endothelin receptor is coupled to phospholipase C via a pertussis toxin-insensitive guanine nucleotide-binding regulatory protein in vascular smooth muscle cells. J Clin Invest 1990;85:653–658.

9 Takuwa Y, Ohue Y, Takuwa N, Yamashita K: Endothelin-1 activates phospholipase C and mobilizes Ca^{2+} from extra- and intracellular pools in osteoblastic cells. Am J Physiol 1989;257:E797–E803.

10 Rodbell M: The role of hormone receptors and GTP-regulatory proteins in membrane transduction. Nature 1980;284:17–22.

11 Rasmussen H, Takuwa Y, Park S: Protein kinase C in the regulation of smooth muscle contraction. FASEB J 1987;1:177–185.

12 Takuwa Y, Kelley G, Takuwa N, Rasmussen H: Protein phosphorylation changes in bovine carotid artery smooth muscle during contraction and relaxation. Mol Cell Endocrinol 1988;60:71–86.

Dr. Yoh Takuwa, Department of Vascular Biology, Faculty of Medicine, University of Tokyo, 7-3-1 Hongo, Tokyo 113 (Japan)

Morii H (ed): Calcium-Regulating Hormones. I. Role in Disease and Aging.
Contrib Nephrol. Basel, Karger, 1991, vol 90, pp 105–110

Effects of the Calcium Channel Antagonist Nicardipine on Renal Action of Endothelin in Dogs

Tokihito Yukimura, Katsuyuki Miura, Yutaka Yamashita,
Tomoji Shimmen, Michiaki Okumura, Shinya Yamanaka,
Michiya Saito, Kenjiro Yamamoto

Department of Pharmacology, Osaka City University Medical School, Osaka, Japan

A lot of evidence has been accumulated to show that the vascular endothelium plays an important role in mediating the actions of many vasoactive substances [1]. Hickey et al. [2] demonstrated that the culture media of the endothelial cells contain a potent vasoconstrictor peptide. Recently, Yanagisawa et al. [3] isolated the vasoconstrictor peptide, endothelin, from the media of cultured porcine endothelial cells and determined its amino acid sequences. Using a synthetic peptide, the vasoconstricting actions were characterized to be dependent on the extracellular calcium ion and antagonized by a calcium channel antagonist, nicardipine, and it has been proposed that endothelin may act as an endogenous agonist for voltage-dependent calcium channels [3]. In our previous study, intrarenal arterial infusion of a synthetic porcine endothelin showed a transient increase in renal blood flow followed by a marked and sustained reduction [4, 5]. Cao and Banks [6] also observed the potent pressor action of the peptide when administered intravenously into the rat, which, however, was not affected by the calcium antagonist verapamil. On the other hand, Hof et al. [7] reported partial contribution of the voltage-dependent calcium channel in endothelin-induced vasoconstriction. The possible involvement of the voltage-dependent calcium channel in vascular actions of endothelin remains controversial. The purpose of this study was to elucidate whether the renal vascular action of the peptide as mediated through activation of the dihydropyridine-sensitive calcium channel and to evaluate the effects of calcium channel antagonists on the responses of glomerular filtration rate and urine formation to the peptide.

Methods

Experiments were performed with adult mongrel dogs of both sexes weighing between 11 and 14 kg. They were anesthetized with sodium pentobarbital intravenously and ventilated artificially by a constant flow respirator. The left kidney was exposed through a retroperitoneal flank incision and renal blood flow was measured with an electromagnetic flowmeter (MFV-2100, Nihon Kohden), as described [8]. Systemic arterial blood pressure was monitored via a catheter placed in the abdominal aorta. A 23-gauge needle was inserted into the left renal artery for the intrarenal arterial administration of saline or endothelin solution at a rate of 0.3 ml/min. Catheters were placed in the brachial artery and left renal vein for blood collection. The left ureter was cannulated for urine collection. The brachial vein was cannulated for an infusion of inulin saline.

After the completion of surgery, animals were left for 1–2 h to allow for stabilization of systematic blood pressure, renal blood flow and urine flow. Urine was then collected during two consecutive 10-min control clearance periods and, at the midpoint of each period, blood samples were taken from the artery and renal vein. After the control periods, a synthetic porcine endothelin (Peptide Institute Inc.) was infused into the renal artery at a rate of 1.0 ng/kg/min in 5 dogs for 25 min. Five minutes after the start of infusion two 10-min urine and blood samples were obtained consecutively.

To determine the renal effects of endothelin in the presence of a voltage-dependent calcium channel antagonist, nicardipine (Yamanouchi Inc.) was infused into the renal artery at a rate of 100 ng/kg/min, and then endothelin was superimposed (1.0 ng/kg/min) in another group of 5 dogs.

Inulin in arterial and renal venous plasma was determined colorimetrically, as described by Walser et al. [9]. Glomerular filtration rate was determined by an arteriovenous difference of plasma inulin concentration. Urinary and plasma concentrations of sodium and calcium were measured by flame photometry (Hitachi 205D) and atomic absorption spectrophotometry (Hitachi, 207), respectively.

The values presented are means $\pm$ SEM. The data were analyzed by one-way analysis of variance between groups or two-way analysis of variance with complete randomized block. Significant differences were determined using the least significant difference test [10].

Results

Intrarenal arterial infusion of a synthetic endothelin in a dose of 1.0 ng/kg/min induced an initial slight increase followed by a sustained decrease in renal blood flow. Twenty minutes infusion of the peptide significantly decreased renal blood flow from the control value of 139 ± 22 to 85 ± 12 ml/min and glomerular filtration rate from 25 ± 3 to 18 ± 2 ml/ min. Urine flow rate and urinary sodium and calcium excretion decreased from 0.59 ± 0.15 to 0.42 ± 0.07 and 59 ± 30 to 34 ± 19 mEq/min and from 0.45 ± 0.21 to 0.17 ± 0.08 mmol/min, 20 min after the start of the peptide infusion, respectively. Although glomerular filtration rate, urine flow and urinary sodium excretion normalized 40 min after cessation of peptide infusion, renal blood flow remained to show a significantly lower value even till 60 min after. The systemic blood pressure did not change during the entire course of the experiment.

Nicardipine infused intrarenally at a dose of 100 ng/kg/min decreased blood pressure by 7 mm Hg below the control values and increase renal blood flow from 168 ± 30 to 198 ± 47 ml/min. Urine flow rate and urinary excretion of calcium increased from the control values, which were statistically insignificant, although urinary excretion of sodium increased from 0.78 ± 0.20 to 1.05 ± 0.17, during intrarenal arterial infusion of nicardipine. Blood pressure and renal blood flow showed that constant values at 10 to 20 min after the start of infusion, and endothelin was then infused (1.0 ng/kg/min) into the renal artery. Renal blood flow remained unchanged 10–15 min after the start of continuous infusion of the peptide and showed a small but significant decrease to 173 ± 43 ml/min following 20 min infusion, with no change in the systemic blood pressure. Glomeruler filtration rate and urine flow rate decreased from 30 ± 6 to 24 ± 4 and 0.95 ± 0.13 to 0.70 ± 0.09 ml/min, following intrarenal administration in dogs superimposed with nicardipine, respectively. Urinary sodium and calcium excretions were also decreased by the peptide administration in dogs superimposed with nicardipine.

The decrease in renal blood flow in dogs treated with nicardipine was $14 \pm 3\%$, which was significantly smaller than that in dogs without treatment ($40 \pm 3\%$), as shown in figure 1. Changes in glomerular filtration rate and fractional excretion of sodium are not affected by nicardipine treatment. The peptide infusion decreased calcium excretion by $50 \pm 14\%$ in control animals and by $49 \pm 6\%$ in nicardipine-treated animals.

Discussion

The present study shows that endothelin is a vasoconstrictor in the dog kidney, since the peptide decreases renal blood flow with no change in the systematic blood pressure. This is consistent with the results obtained by Yanagisawa et al. [3], who first described the amino acid sequences of the peptide and characterized its action on the blood vessels. They proposed that endothelin is an endogenous agonist of the dihydropyridine-sensitive calcium channels, since endothelin-induced constriction in the isolated coronary artery was dependent on the presence of extracellular calcium ion and inhibited by low doses of nicardipine [3]. Kasuya et al. [11] also reported that the peptide promoted calcium influx through voltage-dependent calcium channels and induced vasoconstriction; however, the peptide did not affect the coronary artery binding of dihydropyridine, a calcium antagonist. In the isolated perfused hydronephrotic kidney,

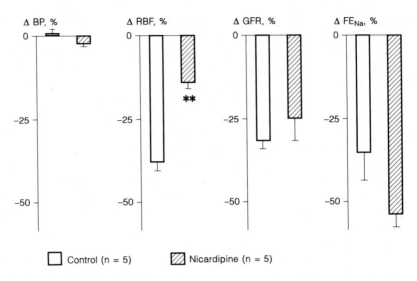

Fig. 1. Changes in systemic blood pressure (BP), renal hemodynamics and sodium excretion in dogs with and without nicardipine treatment. A synthetic porcine endothelin was infused into the left renal artery (i.r.a.) at a dose of 1.0 ng/kg/min and data presented are decreases in percent from the preinfusion values in renal blood flow (RBF), glomerular filtration rate (GFR), and fractional sodium excretion (FE_{Na}). **$p < 0.01$ when comparing the value obtained in dogs without nicardipine (0.1 mg/kg/min, i.r.a.).

Loutzenhiser et al. [12] observed predominant afferent arteriolar vasoconstriction which was completely reversed by nifedipine. On the other hand, Cao and Banks [6] reported that renal vasoconstriction of endothelin did not depend on calcium channels sensitive to verapamil or manganese. Hof et al. [7] failed to observe attenuation of the constrictor effect of endothelin by a calcium antagonist, isradipine, in some vascular beds including kidney in anesthetized rabbits.

It is considered that the differences in the published results about the blocking effects of calcium channel antagonists on peptide-induced vasoconstriction are related to the sensitivity of the type of vessels such as resistance or conductance vessels, or to the different subtypes of endothelin receptors [13]. In the present study, the renal vasoconstriction action of the peptide was clearly attenuated in animals treated with nicardipine and, at least functionally, interacted with dihydropyridine calcium channel blocker, although the precise cellular mechanism by which the renal vasculature responded to the peptide remains unclear.

Endothelin decreased the glomerular filtration rate both in animals treated with and without nicardipine, when administered into the renal artery at a dose of 1.0 ng/kg/min, which was not counteracted by intrarenal administration of nicardipine. We also demonstrated the antidiuretic and antinatriuretic actions of a synthetic porcine endothelin. In our previous study, a smaller dose of endothelin (0.2 ng/kg/min) decreased renal blood flow, urine flow and urinary sodium excretion but not glomerular filtration rate, when given into the renal artery in anesthetized dogs [5]. Fractional sodium excretion was clearly decreased in the present experiment following a higher dose of the peptide (1.0 ng/kg/min), with no change in the systemic circulation. Therefore, it is possible that the peptide may affect water and electrolyte transport in the tubules. Recently, endothelin was reported to stimulate the sodium-hydrogen exchange system, which is involved in maintaining the tonus of vascular smooth muscle and/or enhanced tubular reabsorption [14]. Zeidel et al. [15] reported that endothelin inhibited sodium-pottasium-adenosine triphosphatase in rabbit tubular cells by stimulating prostaglandin production and thereby affecting sodium transport.

In summary, endothelin is a potent vasoconstricting peptide in the kidney and decreases glomerular filtration rate, associated with marked reduction in urine flow rate and urinary sodium and calcium excretion. Reduction of renal blood flow by the peptide was attenuated by a dihydropyridine calcium channel antagonist nicardipine. However, reduction of glomerular filtration rate and antinatriuretic action of the peptide was unaffected by nicardipine treatment. The peptide has a direct effect on glomerular circulation and urine formation. Although the physiological significance of endothelin in regulating renal hemodynamics and urine formation under various conditions has not yet been elucidated, the renal vasoconstrictor action of endothelin is functionally coupled with the activation of the dihydropyridine-sensitive calcium channels.

Summary

Endothelin is a potent vasoconstrictor peptide isolated from cultured vascular endothelial cells. Interaction between endothelin and calcium channel antagonist on the renal hemodynamics and urine formation was studied in anesthetized dogs. Intrarenal arterial administration of the peptide progressively reduced renal blood flow from 139 ± 22 to 85 ± 12 ml/min at 20 min after the start of continuous infusion, with no change in systemic blood pressure. Glomerular filtration rate, urine flow and urinary sodium and calcium excretion decreased significantly by 30–50% from the preinfusion control values. An endothelin-induced reduction in renal blood flow was markedly attenuated by pretreatment with the calcium antagonist nicardipine (100 ng/kg/min intrarenally). Changes in glomerular filtration rate and antinatriuretic and anticalciuretic actions by the peptide was not affected by nicardipine treatment. It is suggested that the renal vasoconstrictor action, but not the tubular action, of endothelin is functionally coupled with the activation of dihydropyridine-sensitive calcium channels.

References

1 Furchgott RF: The role of endothelium in the responses of vascular smooth muscle to drugs. Ann Rev Pharmacol Toxicol 1984;24:175–197.
2 Hickey KA, Rubanyi G, Paul RJ, Highsmith RF: Characterization of a coronary vasoconstrictor produced by cultured endothelial cells. Am J Physiol 1985;248:C550–C556.
3 Yanagisawa M, Kurihara H, Kimura S, Tomobe Y, Kobayashi M, Mitsui Y, Yazaki Y, Goto K, Masaki T: A novel vasoconstrictor peptide produced by vascular endothelial cells. Nature 1988;332:411–415.
4 Miura K, Yukimura T, Yamashita Y, Shimmen T, Okumura M, Imanishi M, Yamamoto K: Endothelin stimulates the renal production of prostaglandin E_2 and I_2 in anesthetized dogs. Eur J Pharmacol 1989;170:91–93.
5 Miura K, Yukimura T, Yamashita Y, Shichino K, Shimmen T, Saito M, Okumura M, Imanishi M, Yamanaka S, Yamamoto K: Effects of endothelin on renal function in anesthetized dogs. Am J Hypertens. 1990;3:632–634.
6 Cao L, Banks RO: Cardivascular and renal actions of endothelin: Effects of calcium-channel blockers. Am J Physiol 1990;258:F254–F258.
7 Hof RP, Hof A, Takiguchi Y: Attenuation of endothelin-induced regional vasoconstriction by isradipine: A nonspecific antivasoconstrictor effect. J Cardiovasc Pharmacol 1990;15(suppl 1):S48–S54.
8 Miura K, Yukimura Y, Imanishi M, Okahara T, Abe Y, Yamamoto K: Effects of SA-446, an angiotensin-converting enzyme inhibitor, on renal function in anesthetized dogs: Special reference to arachidonic acid metabolites. J Cardiovasc Pharmacol 1985;7:102–107.
9 Walser M, Davidson DG, Orloff J: The renal clearance of alkali-stable inulin. J Clin Invest 1955;34:1520–1523.
10 Sokal RS, Rohlf FJ: Biometry, ed 2. New York, Freeman, 1981.
11 Kasuya Y, Ishikawa T, Yanagisawa M, Kimura S, Goto K, Masaki T: Mechanism of contraction to endothelin in isolated porcine coronary artery. Am J Physiol 1989;257:H1828–1835.
12 Loutzenhiser R, Epstein M, Hayashi K, Horton C: Direct visualization of effects of endothelin on the renal microvasculature. Am J Physiol 1990;258:F61–F68.
13 Masuda Y, Miyazaki H, Kondoh M, Watanabe H, Yanagisawa M, Masaki T, Murakami K: Two different forms of endothelin receptors in rat lung. FEBS Let 1989;257:208–210.
14 Richards NT, Piston L, Goldsmith DJA, Cragoe EJ, Hilton PJ: Endothelin-induced contraction of human peripheral resistance vessels is partly dependent on stimulation of sodium-hydrogen exchange. J Hypertens 1989;7:777–780.
15 Zeidel ML, Brady HR, Kone BC, Gullans SR, Brenner BM: Endothelin, a peptide inhibitor of Na^+-K^+-ATPase in intact renal tubular epithelial cells. Am J Physiol 1989;257:C1101–C1107.

Tokihito Yukimura, MD, Department of Pharmacology, Osaka City University
Medical School, 1–4–54, Asahimachi, Abeno-ku, Osaka 545 (Japan)

Morii H (ed): Calcium-Regulating Hormones. I. Role in Disease and Aging.
Contrib Nephrol. Basel, Karger, 1991, vol 90, pp 111–115

Uremic Serum Contains Humoral Factor(s) Larger than Fifty Kilodaltons which Suppresses Endothelin Production in Cultured Endothelial Cells

Hidenori Koyama[a], *Hideki Tahara*[a], *Tetsuo Shoji*[a], *Yoshiki Nishizawa*[a],
Masaaki Inaba[a], *Shuzo Otani*[b], *Masashi Yanagisawa*[c], *Yayoi Ishiguro*[d],
Naoki Takanashi[d], *Hirotoshi Morii*[a]

[a]Second Department of Internal Medicine and [b]Second Department of Biochemistry,
Osaka City University Medical School, Osaka, Japan; [c]Institute of Basic Medical
Sciences, University of Tsukuba, Ibaraki, Japan; [d]Department of Research
Laboratory, SRL Inc., Tokyo, Japan

Endothelin, which is secreted from the endothelial cells, was isolated
and its DNA sequence was determined [1, 3]. Endothelin-1, a major
secretory form of endothelin in human and pig, has multiple actions on the
cardiovascular system including potent vasoconstricting action. Extensive
studies are now under way about the role of this peptide in the pathogenesis of various diseases. Several groups established the radioimmunoassay
system for endothelin by which they revealed an altered plasma level of
endothelin in patients with various conditions [3–7]. We have previously
reported that plasma endothelin concentration is increased in patients with
chronic renal failure in comparison with nonuremic controls [8]. To
investigate the regulatory mechanism for this increase, we examined the
direct effects of uremic serum on the production of endothelin-1 in cultured
endothelial cells.

Material and Methods

Endothelial cells were obtained from porcine aorta enzymatically and cultured in
minimum essential medium containing 10% fetal calf serum at 37 °C in an atmosphere of 5%
CO_2/95% air. Cultures were determined as endothelial cells on the basis of their typical
cobble-stone morphology. Cells beneath the 10th passage were grown in plastic culture dish.
Confluent endothelial cells were washed 3 times with phosphate buffer saline and then
cultured in serum-free medium for 24 h before the experiments.

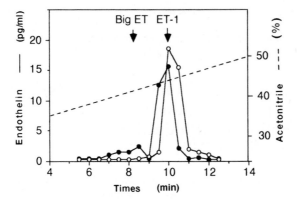

Fig. 1. Profiles of endothelin immunoreactivity in the extracted culture media after treatment with control or uremic serum analyzed by reverse-phase HPLC. ○ = Medium containing control serum; ● = uremic serum.

Radioimmunoassay was done for the determination of endothelin level in culture medium essentially as described previously [8] except that anti-endothelin serum used in this system was obtained from Peninsula's laboratory (California, USA). For the validation of RIA, endothelin immunoreactivity extracted was analyzed on reverse-phase high-performance liquid chromatography (HPLC) using silica ODS column at a flow rate of 1 ml/min. A 0.5-ml aliquot was collected and each fraction was analyzed for its endothelin level with RIA.

Total cellular RNA from porcine endothelial cells was isolated by guanidinium isothiocyanate lysis and ultracentrifugation over cesium chloride cushion. Total RNA (10 μg/lane) was fractionated by electrophoresis in 1% agarose gel containing formaldehyde and blotted onto a nylon membrane (Hybond, N. Amersham). Porcine prepro-endothelin-1 cDNA probe was prepared from a EcoR1 fragment (1.8 kbp) of ppET4-1 and was labelled with [α^{32}P]dCTP by the multiprime DNA labelling system (Amersham). Membranes were hybridized with the labelled probe and exposed to X-ray film for autoradiography. The membranes were subsequently rehybridized with a human interleukin 1-β cDNA.

Results and Discussion

Confluent endothelial cells were cultured in medium containing 10% pooled serum from control subjects and uremic hemodialyzed patients. After treatment of endothelial cells with medium containing control or uremic hemodialyzed serum, endothelin immunoreactivity released into the culture medium was determined with RIA after separation on reverse-phase HPLC. In both media, endothelin immunoreactivity was revealed to be a single peak which comigrated with endothelin-1 (fig. 1). These data indicate that endothelin immunoreactivity released into the culture medium

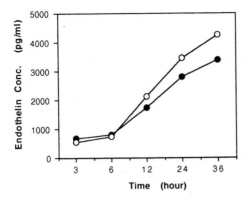

Fig. 2. Endothelin immunoreactivity secreted into the culture medium after the addition of 10% serum-containing medium. Each plot represents the mean immunoreactive endothelin value from duplicate experiments. ○ = Media containing control serum; ● = uremic serum.

consists mostly of intact endothelin-1. Figure 2 shows the levels of endothelin immunoreactivity in the culture medium after treatment with either serum. Endothelin immunoreactivity was decreased in uremic serum culture by 12 h compared with that in the control serum. During these experiments the viability of endothelial cells were not affected by control and uremic serum. We also examined endothelin mRNA levels in endothelial cells after being treated with different sera. Endothelin mRNA levels apparently decreased 3 and 6 h after the addition of medium containing 10% uremic serum compared with control serum. No apparent changes were observed between uremic sera from nondialyzed and hemodialyzed patients. This inhibition was specific and not a result of generalized decrease of mRNAs, since the interleukin 1-β mRNA showed no significant change. For the partial characterization of this inhibitory activity in uremic serum, control and uremic serum were separated into 3 fractions with centricut columns; > 50, 20–50, <20 kD. Each fraction was added to resting endothelial cells, followed by incubation for 3 h. Total RNA was isolated and hybridized with porcine prepro-endothelin-1 cDNA. Fractions larger than 50 kD from uremic serum decreased the level of endothelin-1 mRNA much more than that from control serum, indicating that uremic serum contained inhibitory substance(s) for the endothelin production with molecular weights larger than 50 kD.

From our previous results or others [3–8], plasma endothelin concentrations relatively correlated with blood pressure. However, the plasma level is about 0.1–10 pM, which is much lower than the expected effective concentrations in vitro [1, 9]. This discrepancy suggests that endothelin mainly acts through the local paracrine system, with the plasma level reflecting the amount of local endothelin. In this case, higher concentrations of endothelin in uremic patients represent increased local endothelin production regulated by some unknown factor(s) or decreased catabolism of endothelin in the local system and in plasma. Our results in this study indicate that uremic serum contains substance(s) larger than 50 kD which directly suppresses endothelin production at a gene level in porcine aortic endothelial cells. This inhibitory activity, possibly elevated in chronic renal failure, might play some pathophysiological roles in the regulation of endothelin homeostasis in vascular endothelial cells.

Summary

Direct effects of human uremic serum on the production of endothelin-1 in cultured porcine endothelial cells were examined in this study. Uremic serum decreased the level of monomeric endothelin-1 secreted into the culture medium by endothelial cells. This effect occurred at a transcriptional step because uremic serum decreased the endothelin-1 mRNA level in those cells. For the partial characterization of this inhibitory activity, uremic serum was fractionated with a centricut column. Uremic serum contains humoral factor(s) larger than 50 kD which suppress the endothelin-1 mRNA level in cultured endothelial cells.

References

1   Yanagisawa M, Kurihara H, Kimura S, Tomobe Y, Kobayashi M, Mitsui Y, Yazaki Y, Goto K, Masaki T: A novel potent vasoconstrictor peptide produced by vascular endothelial cells. Nature 1988;332:411–415.
2   Itoh Y, Yanagisawa M, Ohkubo S, Kimura C, Kosaka T, Inoue A, Ishida N, Mitsui Y, Onda H, Fujino M, Masaki T: Cloning and sequence analysis of cDNA encoding the precursor of a human endothelium-derived vasoconstrictor peptide, endothelin: identity of human and porcine endothelin. FEBS Lett 1988;231:440–444.
3   Cernacek P, Steward DJ: Immunoreactive endothelin in human plasma: Marked elevations in patients in cardiogenic shock. Biochem Biophys Res Commun 1989;161:562–567.
4   Miyauchi T, Yanagisawa M, Tomizawa T, Sugishita Y, Suzuki N, Fujino M, Ajisaka R, Goto K, Masaki T: Increased plasma concentrations of endothelin-1 and big endothelin-1 in acute myocardial infarction. Lancet 1989;ii:53–54.
5   Morel DR, Lacroix S, Hemsen A, Steinig DA, Pittet JF, Lundberg JM: Increased plasma and pulmonary lymph levels of endothelin during endotoxin shock. Eur J Pharmacol 1989;167:427–428.
6   Tomita K, Ujiie K, Nakanishi T, Tomura S, Matsuda O, Ando K, Shichiri M, Hirata Y, Marumo F: Plasma endothelin levels in patients with acute renal failure. N Engl J Med 1989;321:1127.

7   Saito Y, Nakao K, Mukoyama M, Imura H: Increased plasma endothelin level in patients with essential hypertension. N Engl J Med 19;322:205.
8   Koyama H, Tabata T, Nishizawa Y, Inoue T, Morii H, Yamaji T: Plasma endothelin levels in patients with uraemia. Lancet 1989;i:991–992.
9   Firth JD, Ratcliffe PJ, Raine AEG, Ledingham JGG: Endothelin: An important factor in acute renal failure? Lancet 1988;ii:1179–1182.

Hidenori Koyama, MD, 2nd Department of Internal Medicine, Osaka City University, Medical School, 1-5-7 Asahi-machi, Abeno-ku, Osaka 545 (Japan)

Calcium and Arteriosclerosis

Morii H (ed): Calcium-Regulating Hormones. I. Role in Disease and Aging.
Contrib Nephrol. Basel, Karger, 1991, vol 90, pp 116–121

Intracellular Signal Transduction Evoked by Low-Density Lipoprotein in Vascular Smooth Muscle Cells

T. Ogihara, R. Morita, S. Morimoto, S. Imanaka, K. Fukuo

Department of Geriatric Medicine, Osaka University Medical School, Osaka, Japan

Low-density lipoprotein (LDL), the major cholesterol-carrying lipoprotein in plasma, is believed to participate in the development of atherogenesis, since it stimulates the proliferation of vascular smooth muscle cells (VSMC) in vitro [1–3] and in vivo [4]. Recently, we [5] and other investigators [6] found that LDL and apolipoprotein-B (Apo-B), the major apoprotein of LDL, induced increases in inositol 1,4,5-trisphosphate (InsP$_3$) and a cytosolic-free Ca^{2+} concentration ([Ca^{2+}]$_i$) in rat cultured VSMC [5, 6]. Moreover, we observed that this lipoprotein also evoked intracellular alkalization of the VSMC [7]. In this study, we investigated possible participation of the signal transduction induced by LDL and Apo-B in the growth-promoting effects in VSMC.

Materials and Methods

LDL and Apo-B were purchases from Sigma. *myo*-[^{3}H]Inositol (98 Ci/mmol) and [^{3}H]thymidine (5 Ci/mmol) were from Amersham. Fura-2 acetoxymethyl ester (fura-2 AM) and 2′,7′-bis(carboxyethyl)carboxy-fluorescein tetra-acetoxymethyl ester (BCECF-AM) were purchased from Dojin Co. (Kumamoto, Japan). Other chemicals used were commercial products of the highest grade available. VSMC were prepared from the thoracic aorta of female Wistar rats by the explant method [8] and cultured in Dulbecco's modified Eagles' medium (DMEM) supplemented with 10% fetal calf serum under 5% CO$_2$ in air. InsP$_3$ was measured by the HPLC method using the anion exchange (SAX) column reported by Irvine et al. [9] with some modification [5]. [Ca^{2+}]$_i$ of the VSMC was measured as described previously [5], measuring the fluorescence of fura 2 with excitation wavelengths of 340 and 380 nm, and an emission wavelength of 495 nm, respectively. [Ca^{2+}]$_i$ was calculated as described by Grynklewicz et al. [10]. The pH$_i$ of the VSMC was measured by the method of Rink et al. [11], measuring the fluorescence of BCECF with excitation wavelengths of 455 and 506 nm and an emission wavelength of 530 nm, respectively [7]. DNA synthesis was measured by the incorporation of [^{3}H]-thymidine as described previously [7].

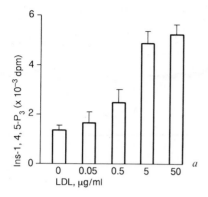

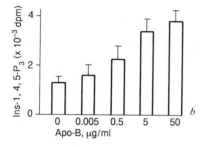

Fig. 1. Dose-dependent changes in production of InsP$_3$ level by LDL (*a*) and Apo-B (*b*) in VSMC. *myo*-[^{3}H]Inositol-labeled VSMC were incubated with different concentrations of LDL or Apo-B for 30 s. Results are means $\pm$ SD for three independent experiments done in triplicate.

Results

LDL (0.5–50 µg/ml) and Apo-B (0.05–5 µg/ml) dose-dependently stimulated InsP$_3$ production in VSMC at 30 s after addition, with maximal accumulations at 50 and 5 µg/ml, respectively (fig. 1). The physiological consequences of enhanced inositol phosphate production were assessed by measuring the changes in [Ca^{2+}]$_i$ induced by LDL and Apo-B. Addition of either LDL at concentrations of more than 0.5 µg/ml or Apo-B at concentrations of more than 0.05 µg/ml caused an increase of [Ca^{2+}]$_i$ within approximately 30 s in the assay solution (fig. 2a). The concentrations of LDL for the half maximal phasic increase in [Ca^{2+}]$_i$ (ED$_{50}$) was 6.2 µg/ml. LDL at concentrations of more than 5 µg/ml caused a prompt but transient decrease in pH$_i$ followed by an increase (fig. 2b). Apo-B at concentrations of more than 0.5 µg/ml also caused prompt but transient

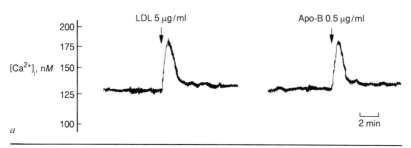

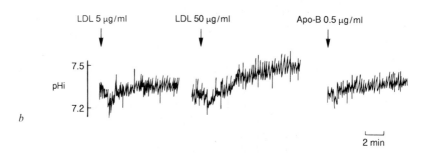

Fig. 2. Changes in $[Ca^{2+}]_i$ *(a)* and in pH_i *(b)* with the additions of LDL (5 µg/ml) and Apo-B (0.5 µg/ml) in VSMC in the control solution. $[Ca^{2+}]_i$ was measured with the fluroescent Ca^{2+} indicator fura 2, and pH_i was measured with the intracellular pH indicator BCECF.

acidification followed by alkalization. Addition of 1 mM amiloride, an inhibitor of the Na^+/H^+ exchanger [12], rapidly reversed the LDL-induced in the pH_i of VSMC. LDL did not cause intracellular alkalization in a Na^+-free solution (data not shown). NH_4Cl caused intracellular alkalization of VSMC without initial transient acidification both in the control solution and in the Na^+-free solution, with a slight increase in the $[Ca^{2+}]_i$ (data not shown).

Moreover, the effect of LDL for 20 h on stimulation of DNA synthesis was dose-dependent with an ED_{50} value of approximately 5.0 µg/ml, almost the same as the ED_{50} for inducing an increase in $[Ca^{2+}]_i$ in VSMC (fig. 3). However, NH_4Cl also stimulated DNA synthesis of VSMC at the same concentrations that induced alkalization of the cells (fig. 3).

Discussion

$InsP_3$ generated from an agonist-stimulated phosphodiesteriatic cleavage of phosphatidylinositol 4,5-bisphosphate is known to cause the release

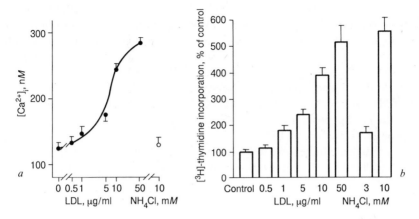

Fig. 3. Effects of LDL and NH$_4$Cl on the $[Ca^{2+}]_i$ (*a*) and on DNA synthesis (*b*) of VSMC. Results are means ± SD for 5 determinations.

of Ca^{2+} from nonmitochondrial stores, especially the endoplasmic reticulum, in a variety of cells including smooth muscle cells [13]. Here we found that InsP$_3$ increased several fold within 30 s after stimulation by either LDL or Apo-B (fig. 1). The changes in the $[Ca^{2+}]_i$ level in the VSMC treated by LDL and Apo-B in our experiment (fig. 2) indicate that changes in inositol phosphates are physiologically significant. It had been suggested that LDL stimulated the DNA syntheses of cultured VSMC from monkeys [3, 14] and from the human splenic artery [2]. In the present experiment, we observed a strong correlation between the phasic rise in $[Ca^{2+}]_i$ and the increase in DNA synthesis by addition of LDL (fig. 3). These results suggest that an increase in $[Ca^{2+}]_i$ may be an important signal for the action of LDL on cell proliferation in VSMC. Involvement of the rise of $[Ca^{2+}]_i$ induced by growth factors in the generation of DNA synthesis has been also reported in VSMC [15].

In this study, we found that LDL and Apo-B also caused intracellular alkalization of VSMC and that this effect was inhibited by amiloride, an inhibitor of the Na$^+$/H$^+$ exchange [12], or by the absence of extracellular Na$^+$. These data indicate that LDL stimulated an amiloride-sensitive Na$^+$/H$^+$ exchange resulting in intracellular alkalization of VSMC. LDL is reported to contain Apo-B, which binds to its receptor on the plasma membrane; so LDL is suggested to stimulate a Na$^+$/H$^+$ exchange through the binding of Apo-B to its receptor. Several growth factors and hormones induce intracellular alkalization mediated by a Na$^+$/H$^+$ exchange, and this alkalization may participate in the initiation of their proliferation [12, 16], but the effect of LDL on the intracellular pH (pH$_i$) is unknown. NH$_4$Cl, which is a permeant weak base [11], also caused an intracellular alkalization, but not through a Na^+/H^+ exchange, because induction of alkalization was not dependent on extracellular Na^+. The transient acidification induced by LDL and Apo-B seemed to be due to the stimulation of Ca^{2+}-ATPase by an increase in $[Ca^{2+}]_i$ [17].

NH_4Cl also stimulated DNA synthesis similar to LDL and Apo-B in VSMC (fig. 3). Our data suggest that intracellular alkalization may participate in the proliferation of VSMC and also partially, at least, in the proliferative effect of LDL.

Further studies are required to clarify the roles and interaction of intracellular Ca^{2+} mobilization and cell alkalization in the induction of the proliferation of VSMC.

Summary

Low-density lipoprotein (LDL) is a well-known causal factor in the development of arteriosclerosis. In the present study, we evaluated LDL-evoked cellular signal transduction in cultured rat vascular smooth muscle cells (VSMC). The addition of LDL at concentrations of more than 50 ng/ml, and apolipoprotein B (Apo-B) at more than 5 ng/ml, induced rapid but transient increases in the inositol 1,4,5-trisphosphate ($InsP_3$) level, and caused rapid phasic and subsequent tonic increases in cytosolic free Ca^{2+} concentration ($[Ca^{2+}]_i$) in a dose-dependent manner in VSMC. LDL and Apo-B also caused transient acidification followed by Na^+-dependent and amiloride-sensitive alkalization of the cells due to stimulation of a Na^+/H^+ exchanger. The enhancement of thymidine incorporation induced by the addition of LDL correlated well with the degree of increment of $[Ca^{2+}]_i$ increases by the lipoprotein. These results suggest that an increase in $[Ca^{2+}]_i$ mediated by $InsP_3$ and intracellular alkalization may function as an important signal for enhanced DNA synthesis induced by LDL in VSMC.

References

1　Goldstein JL, Brown MS: The low-density lipoprotein pathway and its relation to atherosclerosis. Annu Rev Biochem 1977;46:897–930.

2　Oikawa S, Hori S, Sano R, Suzuki N, Fujii Y, Abe R, Goto Y: Effect of low density lipoprotein on DNA synthesis of cultured human arterial smooth muscle cells. Atherosclerosis 1987;64:7–12.

3　Fischer-Dzoga K, Wissler RW: Stimulation of proliferation in stationary primary cultures of monkey aortic smooth muscle cells. Atherosclerosis 1976;24:515–525.

4　Ross R, Glomset JA: The pathogenesis of atherosclerosis. N Engl J Med 1976;295:420–425.

5　Morita R, Morimoto S, Koh E, Fukuo K, Kim S, Itoh K, Taniguchi K, Onishi T, Ogihara T: Low density lipoprotein and apoprotein B induce increase in inositol trisphosphate and cytosolic free Ca^{2+} via pertussis toxin-sensitive GTP-binding protein in vascular smooth muscle cells. Biochem Int 1989;18:647–653.

6   Block LH, Knorr M, Vogt E, Locher R, Vetter W, Groscurth P, Qiao B, Pometta D, James R, Regenass M, Pletscher A: Low density lipoprotein causes general cullular activation with increased phosphatidylinositol turnover and lipoprotein catabolism. Proc Natl Acad Sci USA 1988;85:885–889.

7   Koh E, Morimoto S, Nabata T, Miyashita Y, Kitano S, Morita R, Ogihara T: The action of low density lipoprotein on vascular smooth muscle cells involves increase in intracellular pH. Biochem Int 1990;20:127–133.

8   Ross R: Growth of smooth muscle in culture and formation of elastic fibers. J Cell Biol 1971;50:172–186.

9   Irvine RF, Letcher AJ, Lander DJ, Downes CP: Inositol trisphosphates in carbacol-stimulater rat parotid glands. Biochem J 1984;223:237–243.

10   Grynklewicz G, Poenie M, Tsien RY: A new generation of Ca^{2+} indicators with greatly improved fluorescence properties. J Biol Chem 1985;260:3440–3450.

11   Rink TJ, Tsien RY, Pozzan T: Cytoplasmic pH and free Mg^{2+} in lymphocytes. J Cell Biol 1982;95:189–196.

12   Grinstein S, Rothstein A: Mechanisms of regulation of the Na^+/H^+ exchanger. J Membr Biol 1986;90:1–12.

13   Abdel-Latif AA: Calcium-mobilizing receptors, polyphosphoinositides, and the generation of second messengers. Pharmacol Rev 1986;38:227–272.

14   Yoshida Y, Fischer-Dzoga K, Wissler RW: Effects of normolipidemic high-density lipoproteins on proliferation of monkey aortic smooth muscle cells induced by hyperlipidemic low-density lipoproteins. Exp Mol Pathol 1984;41:258–266.

15   Hirosumi J, Ouchi Y, Watanabe M, Kusunoki J, Nakamura T, Orimo H: Effects of growth factors on cytosolic free calcium concentration and DNA synthesis in cultured rat aortic smooth muscle cells. Tohoku J Exp Med 1989;157:289–300.

16   L'Allemain G, Paris S, Pouyssegur J: Growth factor action and intracellular pH regulation in fibroblasts. J Biol Chem 1984;259:5809–5815.

17   Berk BC, Brock TA, Gimbrone MA Jr, Alexandar RW: Early agonist-mediated ionic events in cultured vascular smooth muscle cells. J Biol Chem 1987;262:5065–5072.

Toshio Ogihara, MD, Department of Geriatric Medicine, Osaka University Medical School, Fukushima-ku, Osaka 553 (Japan)

Calcium and Renal Diseases

Morii H (ed): Calcium-Regulating Hormones. I. Role in Disease and Aging.
Contrib Nephrol. Basel, Karger, 1991, vol 90, pp 124–138

Metabolic and Functional Derangements of Pancreatic Islets in Chronic Renal Failure[1]

Shaul G. Massry, George Z. Fadda

Division of Nephrology and Department of Medicine, University of Southern California School of Medicine, Los Angeles, Calif., USA

Patients with chronic renal failure (CRF) display abnormalities in carbohydrate metabolism [1–5]. They almost always have resistance to the peripheral action of insulin [5, 6], while insulin secretion could be normal [4, 7], increased [8, 9] or decreased [3]. Glucose intolerance is, therefore, usually encountered in uremic patients in whom both impaired tissue sensitivity to insulin and impaired secretion of the hormone coexist [5, 10].

Certain data suggest that parathyroid hormone (PTH) may affect carbohydrate metabolism. Patients with primary hyperparathyroidism may have glucose intolerance [11, 12]. Elevated plasma insulin levels both in the fasting state and in response to glucose [11, 12], as well as insulin resistance [12] have been reported in these patients. It is plausible, therefore, to suggest that the state of secondary hyperparathyroidism which exists in patients with advanced renal failure [13–16] plays an important role in the genesis of the glucose intolerance of uremia. Indeed, studies from our laboratory in dogs [17] have shown that glucose intolerance does not develop with CRF in the absence of PTH and that the hormone does not affect the metabolic clearance of insulin or tissue resistance to insulin in CRF. The normalization of the glucose metabolism in CRF in the absence of excess PTH is most likely due to increased insulin secretion. This latter conclusion was supported by indirect evidence showing that for any given level of blood glucose during intravenous glucose tolerance test, the blood levels of insulin were higher in CRF dogs without excess PTH than CRF animals with secondary hyperparathyroidism (fig. 1). Also, results of studies with hyperglycemic clamp supported this conclusion (fig. 2). These observations indicated, although indirectly, that excess PTH in CRF interferes with insulin secretion.

[1] This work was supported by a grant DK 29955 from the National Institute of Diabetes, Digestive and Kidney Diseases.

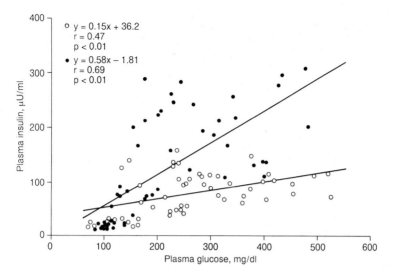

Fig. 1. The relationship between plasma insulin and glucose concentrations observed during intravenous glucose tolerance performed in NPX (○) and NPX-PTX (●) dogs. Reproduced by permission from Akmal et al. [17].

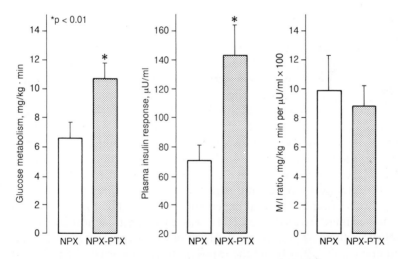

Fig. 2. Glucose metabolism, total insulin response and M/I ratio (total amount of glucose metabolized [M] divided by the total insulin response [1]) observed during the hyperglycemic clamp in NPX and NPX-PTX dogs. Each column represents the mean of data from 6 NPX and 7 NPX-PTX dogs. The brackets denote 1 SE. Asterisks indicate significant difference from NPX with p < 0.01. Reproduced by permission from Akmal et al [17].

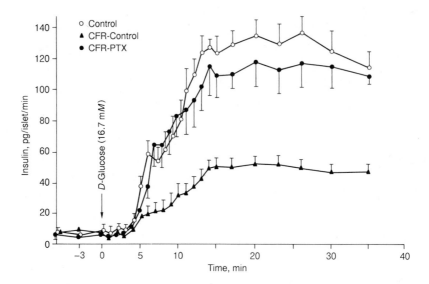

Fig. 3. Dynamic insulin release from perifused pancreatic islets in 5 control rats, 5 CRF animals and 5 CRF-PTX rats. Each data point represents the mean value and bracket 1 SE. Reproduced by permission from Fadda et al. [18].

Direct evidence for such an effect of chronic excess PTH was provided by another study from our laboratory [18]. We found that glucose-induced insulin secretion by pancreatic islets isolated from rats with 6 weeks of CRF is markedly impaired (fig. 3), and this abnormality was absent in normocalcemic parathyroidectomized (PTX) rats with similar duration and degree of CRF. Furthermore, islets isolated from normal rats injected with PTH 1–84 for 6 weeks also displayed a significant impairment in glucose-induced insulin secretion (fig. 4). These data provided evidence that excess PTH in the presence or absence of CRF adversely affects glucose-induced insulin secretion by pancreatic islets.

We also found that total pancreatic calcium content in CRF rats or in normal animals treated with PTH was twice that in normal rats or in CRF-PTX animals [18]. Since PTH is known to enhance entry of calcium into many cells [19–23] and the exposure to chronic excess of PTH with or without CRF is associated with increased calcium content in many tissues [24–30], we proposed that PTH affects pancreatic islets in a similar manner. Indeed, we found [31] that the in vitro exposure of pancreatic islets to PTH is associated with an acute rise in cytosolic calcium ($[Ca^{2+}]_i$)

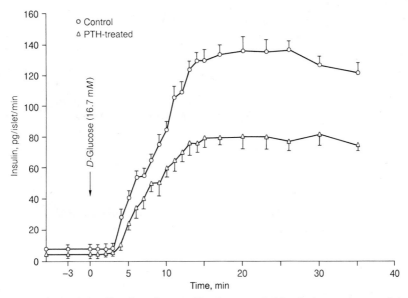

Fig. 4. Dynamic insulin release from perifused pancreatic islets in 4 control rats and 6 PTH-treated rats. Each data point represents the mean value and brackets 1 SE. Reproduced with permission from Fadda et al. [18].

and this effect occurred only when calcium was present in the medium (fig. 5) indicating that PTH augments entry of calcium from the extracellular medium into islets. In addition, the basal levels of $[Ca^{2+}]_i$ in islets from CRF rats (252 ± 7.4 nM) was significantly ($p < 0.01$) higher than those in islets from normal (137 ± 4.5 nM) or in islets from normocalcemic CRF-PTX (158 ± 9.8 nM)rats [32] (fig. 6). Also, chronic treatment of normal rats with PTH (1–84) was associated with marked elevation in the basal levels of $[Ca^{2+}]_i$ in their pancreatic islets (288 ± 27.1) nM) [33] (fig. 7).

Enhanced calcium entry into cells is balanced by pumping calcium out of the cells with Ca^{2+} ATPase, Na^+/Ca^{2+} exchanger and indirectly by Na^+-K^+ ATPase, [34]. Therefore, the high resting level of $[Ca^{2+}]_i$ in islets after chronic exposure to PTH implies that the processes involved in maintaining normal resting levels of cytosolic calcium are not functioning at optimal capacity. We have measured the activity of Ca^{2+} ATPase and found that the V_{max} but not K_m for calcium of this enzyme in islets from CRF rats (8.9 ± 1.26 μmol/mg protein/h) is significantly ($p < 0.01$)lower than in control rats (15.2 ± 0.94 μmol/mg protein/h) or in CRF-PTX animals (12.5 ± 1.13 μmol/mg protein/h) (fig. 8) [33]. The functional integrity of the Ca^{2+} ATPase requires ATP and CRF is associated with

a

124 nM —

137 nM —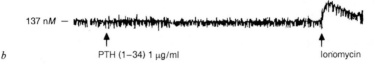

b   PTH (1–34) 1 μg/ml   Ionomycin

256 nM ⌐

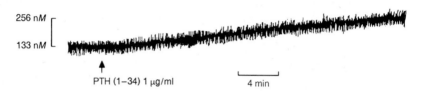

133 nM ⌊

c   PTH (1–34) 1 μg/ml   └─── 4 min ───┘

Fig. 5. Intracellular calcium concentration changes in pancreatic islets in a typical study. Calcium-induced fluorescence of Fura-2 was monitored continuously before and after experimental treatment (arrows). *a* Studies depicting levels of cytosolic calcium in pancreatic islets incubated in a medium containing 1.5 mM calcium and 2.8 mM D-glucose. *b* Studies evaluating cytosolic calcium of pancreatic islets incubated in a medium containing 2.8 mM D-glucose and no calcium before and after the addition of PTH-(1–34) or ionomycin. The latter produced a significant rise in cytosolic calcium due to mobilization of calcium from intracellular stores. *c* Studies evaluating cytosolic calcium in pancreatic islets incubated in a medium containing 1.5 mM calcium and 2.8 mM D-glucose before and after the addition of PTH-(1–34). Reproduced with permission from Fadda et al. [31].

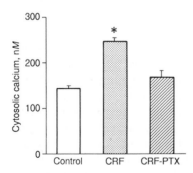

Fig. 6. Resting levels of cytosolic calcium in dispersed cells of pancreatic islets of control, CRF and CRF-PTX rats. Each column represents the mean value of 10–17 measurements provided by islets from 12 rats in each group. Brackets denote SE; *p < 0.01. Data adapted from Fadda et al. [32].

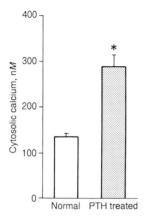

Fig. 7. Resting levels of cytosolic calcium in pancreatic islets from normal rats and PTH-treated animals. Each column represents mean value of 7–10 measurements provided by islets from 6–9 rats; brackets denote 1 SE; *p < 0.01. Reproduced with permission from Perna et al. [35].

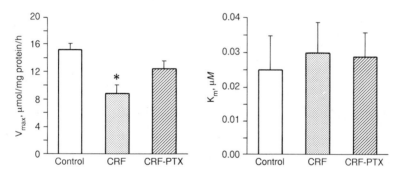

Fig. 8. The V_{max} and K_m of Ca^{2+} ATPase of pancreatic islet membranes of 9 control, 12 CRF and 10 CRF-PTX rats. Each column represents the mean value and the bracket denotes 1 SE. *p < 0.01 vs. control and < 0.05 vs. CRF-PTX. Adapted from Fadda et al. [33].

significant and marked reduction in basal levels of ATP in islets, this derangement is corrected by the PTX of CRF rats (fig. 9). These observations permit the following formulation of the events that lead to sustained elevation in $[Ca^{2+}]_i$ of islets of animals with chronic excess of PTH. The PTH-induced calcium entry into islets would inhibit mitochondrial oxidation and ATP production resulting in low ATP content. The reduction in

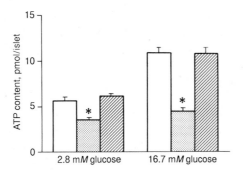

Fig. 9. ATP content of pancreatic islets from control, CRF and CRF-PTX rats. ATP content was estimated after incubation of the islets for 30 min at 37 °C with 2.8 and 16.7 mM D-glucose. Open bars represent mean value in control rats (8–17 studies), dotted bars denote mean value in CRF rats (12–18 studies), and hatched bars show mean value in CRF-PTX rats (15–19 studies). Brackets denote 1 SE. *p < 0.01 vs. control and CRF-PTX rats. Adapted from Fadda et al. [33].

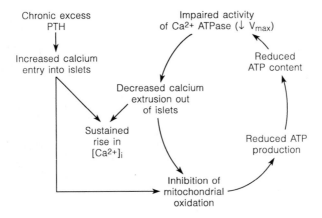

Fig. 10. A schematic presentation of the events leading to a sustained rise in [Ca^{2+}]$_i$ in islets of animals exposed to chronic excess of PTH.

ATP content would impair the activity of Ca^{2+} ATPase leading to decreased calcium extrusion out of islets and hence calcium accumulates in these structure. The rise in [Ca^{2+}]$_i$ would further inhibit mitochondrial oxidation and ATP production, and, thus, a vicious circle develops until a new steady state is achieved with low ATP content, reduced V$_{max}$ of Ca^{2+} ATPase and a sustained rise in [Ca^{2+}]$_i$ of islets (fig. 10).

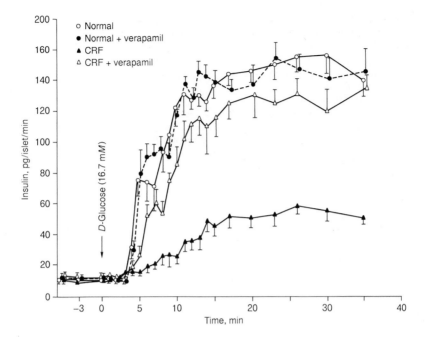

Fig. 11. Dynamic insulin release from perifused pancreatic islets. Each data point depicts mean value and brackets represent 1 SE. Pancreatic calcium was 4.7 ± 0.15, 4.5 ± 0.28, 10.8 ± 0.9 and 5.1 ± 0.44 g/kg dry weight, respectively. Reproduced by permission from Fadda et al. [40].

An increase in the calcium burden of cells may cause derangements in their function. For example, increased calcium content of the myocardium or skeletal muscle is associated with impairment in their bioenergetics [28, 29] and fatty acid oxidation [35, 36]. Increased calcium content of brain synaptosomes results in abnormalities in their norepinephrine metabolism [36] and phospholipid content [38]; and in polymorphonuclear leukocytes is associated with impaired phagocytosis [39]; and of T cells with decreased response to mitogens [40]. It is, therefore, plausible that the PTH-induced elevation in basal levels of $[Ca^{2+}]_i$ of pancreatic islets in CRF is responsible for the impairment in insulin secretion. Studies from our laboratory support this notion in that treatment of CRF rats with the calcium channel blocker, verapamil, from day one of CRF prevented the accumulation of calcium in pancreas and the impairment in glucose-induced insulin secretion by pancreatic islets [41] (fig. 11).

A sustained elevation in basal levels of $[Ca^{2+}]_i$ in CRF may exert several adverse effects on the function and metabolism of the pancreatic

Table 1. Potential mechanisms through which elevated resting levels of [Ca²⁺] of pancreatic islets affects their function and metabolism

1	Impaired glucose uptake
2	Interference with adenylate cyclase-cyclic AMP system
3	Decreased insulin content
4	Reduced availability of ATP
5	Derangements in glucose metabolism of islets
6	Reduced calcium signal, or calcium signal to basal $[Ca^{2+}]_i$ ratio
7	Disturbances in protein kinase C and/or calmodulin system

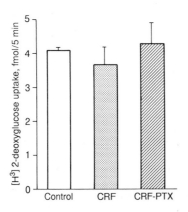

Fig. 12. [H³]2-deoxyglucose uptake by pancreatic islets from control, CRF and CRF-PTX rats. Each column represents the mean values of 9 measurements in control, 5 in CRF and 9 CRF-PTX rats. Brackets denote 1 SE. There were no significant differences between the three groups of animals. Adapted from Fadda et al. [33].

islets (table 1). One or more of these potential derangements may, therefore, underlie the impairment in insulin secretion in CRF. We found [35] that glucose uptake by (fig. 12) and cyclic AMP production of (fig. 13) pancreatic islets from CRF rats are not impaired. Insulin content of islets from CRF rats are significantly ($p < 0.01$) lower than that in islets from normal or CRF-PTX animals (fig. 14) [35]. The low insulin content of these islets may have contributed to the impairment in insulin release. However, one must be cautious in accepting this interpretation since the content of insulin in islets of CRF rats was 30 times higher than the amount secreted during the 30 min of the study. Further, *D*-glyceraldehyde caused a normal secretion of insulin from the islets of CRF rats despite their lower insulin content (fig. 15) [33].

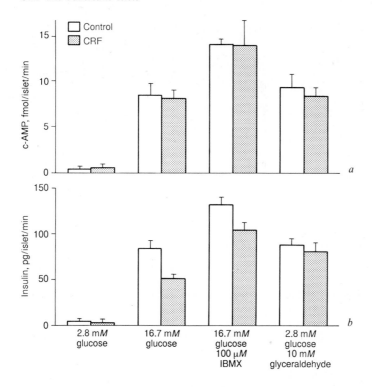

Fig. 13. Static studies depicting the effects of D-glucose, IBMX and glyceraldehyde on both cyclic AMP content of (a) and insulin release (b) from pancreatic islets of control and CRF rats. Each column represents mean values of 4 rats and brackets denote 1 SE.

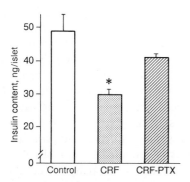

Fig. 14. Insulin content of pancreatic islets from control, CRF and CRF-PTX rats. Each column represents the mean value of 23 measurements in control, 22 in CRF and 15 in CRF-PTX rats. Brackets denote 1 SE. *p < 0.01 vs. control and CRF-PTX rats. Adapted from Fadda et al. [33].

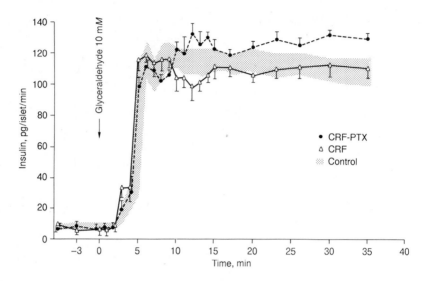

Fig. 15. Dynamic studies of D-glyceraldehyde-induced insulin release from perifused pancreatic islets in 4 control, 4 CRF, and 4 CRF-PTX rats. Insulin secretion was first measured during dynamic perfusion with KRB containing 2.8 mM D-glucose. At 0 time, the perifusate was changed to KRB containing 2.8 mM D-glucose and 10 mM D-glyceraldehyde. The shaded area represents mean ± 1 SE. Each data point represents the mean value and brackets denote ± 1 SE. There was no significant difference between the insulin secretion in the three groups of animals. Reproduced by permission from Fadda et al. [33].

Several observations reported by us [33] are consistent with impaired glucose metabolism by the islets of CRF rats. After entry of glucose into the islets, it is phosphorylated to glucose-6-phosphate, converted by an isomerase to fructose-6-phosphate, and further phosphorylated by phos-phofructokinase-1 (PFK-1) to fructose 1,6-biphosphate [42]. These processes require adequate amounts of ATP and intact function of the PFK-1. Our findings of low basal ATP (fig. 9) content and noncompetitive inhibition of PFK-1 (normal K_m and reduced V_{max}) (fig. 16) indicate that glucose metabolism is impaired at a step(s) of the glycolytic pathway before the production of glyceraldehyde-3-phosphate. Further support for impaired glycolysis in islets from CRF rats is provided by our observations that net lactic acid production in these islets is markedly lower than in those from control or CRF-PTX rats (fig. 17).

In addition, our demonstration that glyceraldehyde-induced insulin release by islets from CRF rats is not different from that of islets from normal rats (fig. 15) are also consistent with the interpretation that CRF

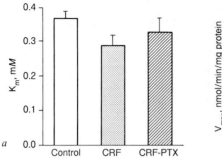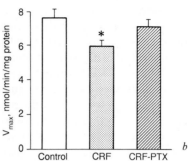

Fig. 16. The K_m for fructose-6-phosphate (a) and V_{max} (b) for PFK-1 measured in islets from control (9 studies), CRF (7 studies), and CRF-PTX (3 studies). Each study required 2 rats. Each column represents the mean values and brackets denote 1 SE. *$p < 0.01$ vs. control and CRF-PTX. Adapted from Fadda et al. [33].

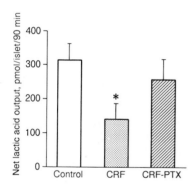

Fig. 17. Net lactic acid output by islets from 17 control rats, 11 CRF animals, and 10 CRF-PTX rats. Each bar represents the mean value and the brackets 1 SE. *$p < 0.01$ vs. control and CRF-PTX rats. Adapted from Fadda et al. [33].

with excess PTH impairs glucose metabolism in islets, since glyceraldehyde enters the glycolytic pathway at the triose phosphate isomerase levels [42].

Calcium serves as a second messenger in many biological functions of the cell [43]. In order for calcium to act as such, the ratio between the calcium signal (rise in $[Ca^{2+}]_i$ induced by an agonist) and the background (resting $[Ca^{2+}]_i$) should be of certain magnitude and adequately large [44]. Therefore, in any situation where the resting levels of cytosolic calcium are elevated, the signal to background ratio will be smaller than in conditions with normal resting cytosolic calcium, and therefore the biological response to the calcium signal will be reduced. Thus, it is reasonable to suggest that the elevated levels of resting cytosolic calcium in islets of CRF rats would be associated with reduced glucose-induced insulin release.

The low ATP content may affect insulin secretion through another pathway besides its effect on glucose metabolism. Available data assign an important role for ATP in the process of insulin secretion by the pancreatic islets. ATP facilitates the closure of ATP-dependent potassium channels [45, 46], which is followed by cell depolarization [46, 47] and subsequent activation of voltage-sensitive calcium channels [46, 48]. As a consequence, calcium enters the islets causing a rise in cytosolic calcium concentration that triggers cellular events that lead to insulin secretion. Corkey et al. [49] postulated that it is the ATP/ADP ratio that is important in the sequence of events described above, and a rise in the ATP/ADP ratio initiates the closure of the ATP-sensitive potassium channels and the depolarization of islets. Thus, the lower ATP content and/or the lower ATP/ADP ratio in islets of CRF rats [33] both in the resting state and after exposure to 16.7 mM D-glucose (fig. 9) may contribute to the impaired insulin secretion through an effect on the ATP-dependent potassium channels [45, 46].

References

1   Neubauer E: Über Hyperglykämie bei Hochdrucknephritis und die Beziehungen zwischen Glykämie und Glykosurie bei Diabetes mellitus. Biochem Z 1910;25:284–295.
2   Westervelt FG, Schreiner GE: The carbohydrate intolerance of uremic patients. Ann Intern Med 1962;57:266–275.
3   Hampers CL, Soeldoner JS, Doak PB, Merrill JP: Effect of chronic renal failure and hemodialysis on carbohydrate metabolism. J Clin Invest 1966;45:1719–1931.
4   Horton ES, Johnson C, Lebovitz HE: Carbohydrate metabolism in uremia. Ann Intern Med 1968;68:63–74.
5   DeFronzo RA, Andres R, Edgar P, Walker WG: Carbohydrate metabolism in uremia. A review. Medicine 1973;52:469–481.
6   DeFronzo RA, Alverstrand A, Smith D, Hendler R, Hendler E, Wahren J: Insulin resistance in uremia. J Clin Invest 1981;67:563–568.
7   Samaan NA, Freeman RM: Growth hormone levels in severe renal failure. Metabolism 1970;19:102–113.
8   Lowrie EG, Soeldner JS, Hampers CL, Merrill JP: Glucose metabolism and insulin secretion in uremic, prediabetic in normal subjects. J Lab Clin Med 1970;76:603–615.
9   Hutching RH, Hagstron RM, Scribner BH: Glucose intolerance in patients on long-term intermittent dialysis. Ann Intern Med 1966;65:275–285.
10   DeFronzo RA: Pathogenesis of glucose intolerance in uremia. Metabolism 1978;27:1866–1880.
11   Ginsberg H, Olefsky JM, Reaven GM: Evaluation of insulin resistance in patients with primary hyperparathyroidism. Proc Exp Biol Med 1975;148:942–945.

12   Kim H, Kalkhoff RK, Costrini NV, Cerletty JM, Jacobson M: Plasma insulin distur-
     bances in primary hyperparathyroidism. J Clin Invest 1971;50:2596–2605.
13   Pappenheimer AM, Wilens SL: Enlargement of the parathyroid glands in renal disease.
     Am J Pathol 1935;11:73–91.
14   Roth SI, Marshall RB: Pathology and ultrastructure of the human parathyroid glands
     in chronic renal failure. Arch Intern Med 1969;124:390–407.
15   Berson SA, Yalow R: Parathyroid hormone in plasma in adenomatous hyperparathy-
     roidism, uremia and bronchogenic carcinoma. Science 1968;154:907–909.
16   Massry SG, Coburn JW, Peacock M, Kleeman CR: Turnover of endogenous parathy-
     roid hormone in uremic patients and those undergoing hemodialysis. Trans Am Soc
     Artif Intern Organs 1972;8:422–426.
17   Akmal M, Massry SG, Goldstein AD, Fanti P, Weisz A, DeFronzo RA: Role of
     parathyroid hormone in the glucose intolerance of chronic renal failure. J Clin Invest
     1985;75:1037–1044.
18   Fadda GZ, Akmal M, Premdas FH, Lipson LG, Massry SG: Insulin release from
     pancreatic islets. Effect of CRF and excess PTH. Kidney Int 1988;33:1066–1072.
19   Chausmer AB, Sherman BS, Wallach S: The effect of parathyroid hormone on hepatic
     cell transport of calcium. Endocrinology 1972;90:663–672.
20   Borle AB: Calcium metabolism at the cellular level. Fed Proc 1973;30:1944–1950.
21   Bogin E, Massry SG, Harary I: Effect of parathyroid hormone on rat heart cells. J Clin
     Invest 1981;67:1215–1227.
22   Bogin E, Massry SG, Levi J, Djaldeti M, Bristo G, Smith J: Effect of parathyroid
     hormone on osmotic fragility of human erythrocyte. J Clin Invest 1983;69:1017–1025.
23   Fraser CL, Sarnacki P, Budayr A: Evidence that parathyroid hormone-mediated cal-
     cium transport in rat brain synaptosomes is independent of cyclic adenosine monophos-
     phate. J Clin Invest 1988;81:982–988.
24   Berkow JW, Fine BS, Zimmerman LE: Unusual occular calcification in hyperparathy-
     roidism. Am J Opthalmol 1986;66:812–824.
25   Massry SG, Coburn JW, Hartenbower DL, Shinaberger JH, DePalma JR, Chapman E,
     Kleeman CR: Mineral content of human skin in uremia: Effect of secondary hyper-
     parathyroidism and hemodialysis. Proc Eur Dialy Transplant Assoc 1970;7:146–150.
26   Bernstein DS, Pletka P, Hattner RS, Hampers CL, Merril JP: Effect of total parathy-
     roidectomy and uremia on the chemical composition of bone, skin, and aorta in the rat.
     Israel J Med Sci 1971;7:513–514.
27   Kraikipanitch S, Lindeman RD, Yoenice AA, Baxter DJ, Haygood CC, Blue MM:
     Effect of azotemia and myocardial accumulation of calcium. Miner Electrolyte Metab
     1978;1:12–20.
28   Akmal M, Goldstein DA, Multani S, Massry SG: Role of uremia, brain calcium and
     parathyroid hormone on changes in electroencephalogram in chronic renal failure. Am
     J Physiol 1984;246:F575–579.
29   Baczynski R, Massry SG, Kohan R, Magott M, Saglikes Y, Brautbar N: Effects of
     parathyroid hormone on myocardial energy metabolism in the rat. Kidney Int
     1984;27:718–725.
30   Bacynski R, Massry SG, Magott M, El-Belbessi S, Kohan R, Brautbar N: Effect of
     parathyroid hormone on energy metabolism of skeletal muscle. Kidney Int 1985;28:722–
     727.
31   Akmal M, Massry SG: Role of parathyroid hormone in the decreased motor nerve
     conduction velocity of chronic renal failure. Proc Soc Exp Biol Med 1990;195:202–207.
32   Fadda GZ, Akmal M, Lipson LG, Massry SG: Direct effect of parathyroid hormone on
     insulin secretion from pancreatic islets. Am J Physiol 1990;258:E975–E984.

33 Fadda GZ, Hajjar SM, Perna AF, Zhou X-J, Lipson LG, Massry SG: On the mechanism of impaired insulin secretion in chronic renal failure. J Clin Invest 1991;87:255–261.

34 Carafoli E, Crompton M: The regulation of intracellular calcium. Curr Top Membr Trans 1987;10:151–216.

35 Perna AF, Fadda GZ, Zhou X-J, Massry SG: Mechanisms of impaired insulin secretion after chronic excess of parathyroid hormone. Am J Physiol 1990;259:F210–F216.

36 Smogorzewski M, Perna AF, Borum PR, Massry SG: Fatty acid oxidation in the myocardium effect of parathyroid hormone and CRF. Kidney Int 1988;34:797–803.

37 Smogorzewski M, Piskorska G, Borum PR, Massry SG: Chronic renal failure, parathyroid hormone and fatty acids oxidation in skeletal muscle. Kidney Int 1988;33:555–560.

38 Islam A, Smogorzewski M, Massry SG: Effect of chronic renal failure and parathyroid hormone on phospholipid content of brain synaptosomes. Am J Physiol 1989;256:F705–F710.

39 Smogorzewski M, Campese VM, Massry SG: Abnormal norepinephrine uptake and release in brain synaptosomes in chronic renal failure. Kidney Int 1989;36:458–465.

40 Alexiewicz JM, Smogorzewski M, Fadda GZ, Massry SG: Impaired phagocytosis (PHAGO) in dialysis patients (DP): Studies on mechanism. Proc Am Soc Nephrol 1990;23:A in press.

41 Fadda GZ, Akmal M, Soliman R, Lipson LG, Massry SG: Correction of glucose intolerance and the impaired insulin release of chronic renal failure by verapamil. Kidney Int 1989;36:773–779.

42 Stadtman ER: Allosteric regulation of enzyme activity. Adv Enzymol Relat Areas Mol Biol 1966;28:141–154.

43 Rasmussen H, Barrett PQ: Calcium messenger system: An integrated view. Physiol Rev 1984;64:938–984.

44 Blaustein M: Calcium transport and buffering in neurones. Trends Neurosci 1988;11:438–443.

45 Cook DL, Hales CN: Intracellular ATP directly blocks K^+ channels in pancreatic β-cells. Nature Lond 1984;311:271–273.

46 Prentki M, Matchinsky FM: Ca^{2+}, cAMP and phospholipid derived messengers in coupling mechanisms of insulin secretion. Physiol Rev 1987;67:1185–1248.

47 Henquin JC, Meissner HP: Significance of ionic fluxes and changes in membrane potential for stimulus secretion coupling in pancreatic β-cells. Experientia 1984;40:1043–1052.

48 Hedeskov CJ: Mechanisms of glucose-induced insulin secretion. Physiol Rev 1980;60:442–509.

49 Corkey BE, Deeney JT, Glennon MC, Matschinsky FM, Prentki M: Regulation of steady-state free Ca^{2+} levels by ATP/ADP ratio and orthophosphate in permeabilized RINm5F insulinoma cells. J Biol Chem 1988;263:4247–4253.

Shaul G. Massry, MD, Los Angeles County – USC Medical Center,
Division of Nephrology, 1200 North State Street, Room 4250, Los Angeles,
CA 90033 (USA)

Vitamin D Metabolism and Other Parameters in Renal Diseases

Morii H (ed): Calcium-Regulating Hormones. I. Role in Disease and Aging.
Contrib Nephrol. Basel, Karger, 1991, vol 90, pp 139–143

Vitamin D Metabolism in Nephrotic Rats

Masayasu Mizokuchi, Minoru Kubota, Yasuhiko Tomino, Hikaru Koide

Division of Nephrology, Department of Medicine, Juntendo University School of Medicine, Tokyo, Japan

Hypocalcemia, reduced intestinal absorption of calcium, osteomalacia and hyperparathyroidism have been demonstrated in patients with nephrotic syndrome and normal renal function [1, 2]. Reduced serum levels of vitamin D metabolites, 25(OH)D, 1,25(OH)$_2$D and 24,25(OH)$_2$D are found in many nephrotic patients [2–4]. These low serum levels of vitamin D metabolites are presumably due to its urinary loss with vitamin D-binding protein [3, 5]. But, this fact cannot explain all the abnormalities of calcium and vitamin D metabolism in nephrotic syndrome. To evaluate other mechanisms of abnormal calcium and vitamin D metabolism in nephrotic syndrome, we studied the serum levels of vitamin D metabolites, the kinetics of renal 25(OH) vitamin D-1-hydroxylase activity in vitro and nephrogeneous cyclic AMP in regard to response to exogenous PTH administration in puromycin aminonucleoside (PAN)-induced nephrotic rats.

Materials and Methods

Male Sprague-Dawley rats weighing 250 g were given subcutaneous injections of 1.5 mg/ 100 g body weight PAN for 12 days. Some of these rats were given intraperitoneal injection of 100 IU of 25(OH)D$_3$ for last 3 days and intraperitoneal injection of 10 IU of human PTH 1–34 1 h prior to use. We measured serum vitamin D metabolites, 25(OH)D, 1,25(OH)$_2$D, mid-molecule PTH, renal 25(OH)D-1-hydroxylase in vitro [6], and nephrogenous cyclic AMP before and after PTH injection.

Results

The serum biochemical data of the various groups of rats, control, nephrotic and 25(OH)D$_3$-injected nephrotic, are presented in table 1. The nephrotic and 25(OH)D$_3$-injected nephrotic rats had heavy proteinuria, but

Table 1. Serum biochemical data (mean ± SEM)

Group	Ca mg/dl	P mg/dl	Mg mEq/l	Creatinine mg/dl	Urine protein mg/24 h
Control	9.97 ± 0.40	7.48 ± 0.55	2.20 ± 0.20	0.79 ± 0.20	7.1 ± 1.1
Nephrotic	5.26 ± 0.55[a]	7.76 ± 0.85	2.20 ± 0.15	0.85 ± 0.20	215.2 ± 39.3[a]
Nephrotic +25(OH)D$_3$	4.49 ± 0.28[a]	7.74 ± 0.60	2.23 ± 0.17	0.83 ± 0.36	2.06 ± 30.1[a]

[a] Significant from controls, p < 0.01.

Table 2. Kinetics of 1α-hydroxylase (mean ± SE)

Group	V$_{max}$ ng/300 mg tissue/20 min	Apparent K$_m$ × 10^{-5} M
Control	56.9 ± 4.6	1.9 ± 0.7
Nephrotic	28.5 ± 2.2[a]	1.0 ± 0.4

[a] Significant from controls (p < 0.01).

Table 3. Nephrogenous cyclic AMP (nmol/100 ml GF; mean ± SE)

Group	Before PTH injection	After PTH injection
Control	2,303.7 ± 266.4	28,809.1 ± 4,158.2
Nephrotic	282.2 ± 32.6[a]	413.0 ± 78.5

[a] The value is significant from controls (p < 0.01).

the serum creatinine of these groups remained normal. Hypocalcemia was found in nephrotic and 25(OH)D$_3$-injected nephrotic rats, but there were no differences in the serum concentrations of phosphate and magnesium.

The plasma concentration of ionized calcium in control rats (1.71 ± 0.07 mEq/l) was significantly higher than that observed in nephrotic rats (0.86 ± 0.16 mEq/l) and 25(OH)D$_3$-injected nephrotic rats (1.39 ± 0.09 mEg/l). The concentration of ionized calcium in nephrotic rats was significantly lower than in the 25(OH)D$_3$-injected nephrotic rats.

The serum levels of PTH in nephrotic rats (242.5 ± 9.3 pmol/l) and 25(OH)D$_3$-injected nephrotic rats (255.1 ± 18.6 pmol/l) were significantly

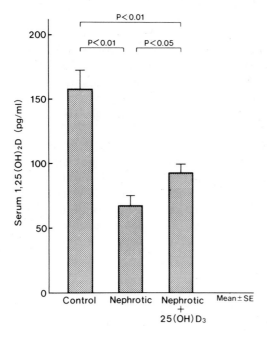

Fig. 1. Serum 1,25(OH)$_2$D level.

higher than in the control rats (144.5 ± 6.2 pmol/l), but there were no differences between nephrotic rats and 25(OH)D$_3$-injected rats.

The serum level of 25(OH)D in 25(OH)D$_3$-injected nephrotic rats (24.7 ± 2.8 ng/ml) was significantly higher than in control rats (10.9 ± 1.2 ng/ml). The serum level of 25(OH)$_2$D in nephrotic rats was not detectable in our assay system.

The serum levels of 1,25(OH)$_2$D in three groups are shown in figure 1. The serum level of 1,25(OH)$_2$D in nephrotic rats was significantly lower than in control rats. The serum level of 1,25(OH)$_2$D in 25(OH)D$_3$-injected nephrotic rats was significantly higher than in nephrotic rats, but was significantly lower than in control rats.

The kinetics of renal 25(OH)D-1-hydroxylase in vitro are presented in table 2. Despite the elevation of serum level of PTH, V$_{max}$ of renal 25(OH)D-1-hydroxylase in nephrotic rats was lower than in control rats.

The response of nephrogenous cyclic AMP to human PTH 1–34 in control and nephrotic rats is shown in table 3. The basal level of nephrogenous cyclic AMP in nephrotic rats was significantly lower than in control rats. And the response of nephrogenous cyclic AMP to PTH in nephrotic rats was also lower than in control rats.

Discussion

The serum level of 25(OH)D in 25(OH)D$_3$-injected nephrotic rats was significantly higher than in control rats, and the serum level of 1,25(OH)$_2$D in these rats was significantly lower than in the control rats, but significantly higher than in nephrotic rats. This fact suggests that the low concentration of serum 1,25(OH)$_2$D in nephrotic rats might be partially due to the low concentration of substrate (25(OH)D).

Despite the elevation of the serum level of PTH, V_{max} of renal 25(OH)D-1-hydroxylase in nephrotic rats was lower than in control rats, and the response of nephrogenous cyclic AMP to PTH was impaired. These facts suggest that abnormalities in calcium and vitamin D metabolism in nephrotic syndrome might be attributed to both urinary loss of vitamin D metabolites and impaired proximal tubular function.

Summary

It is well known that patients with nephrotic syndrome and normal renal function have hypocalcemia in spite of high PTH concentration, caused by the low serum concentration of the active vitamin D metabolite, 1,25(OH)$_2$D, presumably due to its loss in urine. However, it has been uncertain whether the conversion of 25(OH)D into 1,25(OH)$_2$D in the kidney is impaired. In this study, we examined the responsibility of 1,25(OH)$_2$D in PAN-induced nephrotic rats. Sprague-Dawley rats weighing 250 g were given subcutaneous injections 1.5 mg/100 g PAN for 12 days prior to use. Some of these rats were given intraperitoneal injection of 100 IU of 25(OH)D$_3$ for 3 days prior to use and of 10 IU of PTH. We measured Ca^{2+} in plasma, vitamin D metabolites and mid-molecule PTH in serum, renal 25(OH)D-1-hydroxylase activity in vitro, and response of nephrogenous cyclic AMP to exogenous PTH administration. In nephrotic rats, plasma Ca^{2+}, serum 25(OH)D and 1,25(OH)$_2$D were lower than in control rats, and the serum PTH level was higher than in controls. In 25(OH)D$_3$-injected nephrotic rats, Ca^{2+} and 1,25(OH)$_2$D were higher than in nephrotic rats, indicating that the decreased level of 1,25(OH)$_2$D in nephrotic rats was partially due to the low serum level of 25(OH)D. Despite the elevation of the serum level of PTH, the V_{max} of renal 25(OH)D-1-hydroxylase in nephrotic rats was lower than in controls. Response of nephrogenous cyclic AMP to PTH in nephrotic rats was lower than in controls. Although nephrotic rats had higher PTH levels than control rats, V_{max} of renal 25(OH)D-1-hydroxylase and response of cyclic AMP to exogenous PTH administration in nephrotic rats were lower than in controls, suggesting that abnormalities of calcium metabolism in patients with nephrotic syndrome might be partially attributed to the impaired renal response to PTH.

References

1   Goldstein DA, Haldman B, Sherman D, Norman AW, Massry SG: Vitamin D metabolites and calcium metabolism in patients with nephrotic syndrome and normal renal function. J Clin Endocrinol Metabol 1981;52:116–121.

2   Malluche HH, Goldstein DA, Massry SG: Osteomalacia and hyperparathyroid bone
    disease in patients with nephrotic syndrome. J Clin Invest 1979;63:494–500.
3   Lambert PW, De Oreo PB, Fu IY, Kaetzel DM, von Ahn K, Hollis BW, Roos BA:
    Urinary and plasma vitamin D_3 metabolites in the nephrotic syndrome. Metab Bone Dis
    Rel Res 1982;4:7–15.
4   Auwerx J, De Keyser L, Bouillon R, De Moor P: Decreased free 1,25-dihydroxychole-
    calciferol index in patient with the nephrotic syndrome. Nephron 1986;42:231–235.
5   Barragry JM, France MW, Carter ND, Auton JA, Beer N, Boucher BJ, Cohen RD:
    Vitamin-D metabolism in nephrotic syndrome. Lancet 1977;ii:629–632.
6   Horiuchi N, Shinki T, Suda S, Takahashi N, Yamada S, Takayama H, Suda T: A rapid
    and sensitive in vitro assay of 25-hydroxyvitamin D_3-1-hydroxylase and 24-hydroxylase
    using rat kidney homogenates. Biochem Biophys Res Commun 1984;121:174–180.

Dr. Masayasu Mizokuchi, Division of Nephrology, Department of Medicine,
Juntendo University School of Medicine, Hongo 2-1-1 Bunkyo-ku,
Tokyo 113 (Japan)

Morii H (ed): Calcium-Regulating Hormones. I. Role in Disease and Aging.
Contrib Nephrol. Basel, Karger, 1991, vol 90, pp 144–146

Circulating Levels of Vitamin D Metabolites after Renal Transplantation

Shigeo Nakajima[a], *Kanji Yamaoka*[a], *Hiroyuki Tanaka*[b], *Yoshiki Seino*[b]

[a]Department of Pediatrics, Osaka University Medical School, Osaka, Japan;
[b]Department of Pediatrics, Okayama University School of Medicine, Okayama, Japan

Renal osteodystrophy is one of the most important complications in patients with chronic renal failure, and disturbed metabolisms of 1,25-dihydroxyvitamin D_3 (1,25-$(OH)_2D_3$) and parathyroid hormone (PTH) are considered to be the main factors responsible for this disorder. Because 1,25-$(OH)_2D_3$ and 24,25-$(OH)_2D_3$ are converted from 25-OHD_3 in the renal tubular cells, the change in the serum levels of vitamin D metabolites after renal transplantation is of much interest. Although their serum levels several months or years after the renal transplantation have been reported, little is known about the rapid change in vitamin D synthesis in the transplanted kidney after the transplantation. In this study, we measured the serum levels of vitamin D metabolites during the first few weeks after renal transplantation.

Materials and Methods

We measured serum 25-OHD, 1,25-$(OH)_2D$ and 24,25-$(OH)_2D$ in 5 uremic children who underwent renal transplantation at Osaka University Hospital. Three of them had congenital hypoplastic kidney, one had reflux nephropathy and one had primary hyperoxaluria. All the patients were administered prednisolone, azathioprine and ciclosporin as immunosuppressive agents after transplantation. Blood samples were taken on the 0, 1st, 2nd and 3rd days and, after that, once a week for several weeks. Serum concentrations of 25-OHD and 24,25-$(OH)_2D$ were determined by the competitive protein-binding assay [1] and 1,25-$(OH)_2D$ was measured by the radio-receptor binding assay [2], as previously reported.

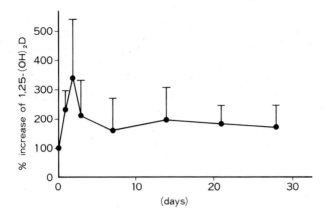

Fig. 1. Change of serum 1,25-(OH)$_2$D concentration after renal transplantation in 5 uremic patients. Data are expressed as mean ± SD.

Results

In all patients, the serum 1,25-(OH)$_2$D level showed a transient increase on the 1st or 2nd day after the operation (fig. 1). In 2 cases, it remained within normal levels after a slight decline. In the other 3 patients, it decreased after the initial peak and increased gradually to the normal level again within 2–4 weeks. Serum 25-OHD level remained low for 4–8 weeks after renal transplantation. Serum 24,25-(OH)$_2$D concentration was also very low after transplantation, but recovered to normal after 1,25-(OH)$_2$D became normal.

Discussion

Previous studies regarding serum 1,25-(OH)$_2$D concentration after renal transplantation have given contradictory results. Lund et al. [3] and Sakhaee et al. [4] reported that it decreased and Riancho et al. [5] demonstrated a positive correlation between serum 1,25-(OH)$_2$D level and GFR. However, Garabedian et al. [6] reported an extremely high concentration of 1,25-(OH)$_2$D at the 3rd to 6th months after transplantation. In our study, which was focused on the rapid change in serum vitamin D metabolite levels after renal transplantation, we observed an initial rise in serum 1,25-(OH)$_2$D concentration on the 1st or 2nd day, and 2–4 weeks were required for the 1,25-(OH)$_2$D levels to return to normal.

Several factors, such as PTH, calcitonin, phosphorus and calcium are known to control serum 1,25-(OH)$_2$D concentration. Because these factors affect each other, it is difficult to determine the cause of the initial peak of 1,25-$(OH)_2D$. Previously, hyperparathyroidism and hypophosphatemia have been revealed to exist in transplanted patients [7, 8], and both of these factors are known to stimulate the production of 1,25-$(OH)_2D$ in the kidney. Moreover, immunosuppressive agents, such as prednisolone and ciclosporin, may possibly influence renal tubular 25-OHD-1α-hydroxylase activity.

Summary

We investigated the change in the serum levels of vitamin D metabolites after renal transplantation in 5 uremic children. The serum 1,25-$(OH)_2D$ level showed a transient increase on the 1st or 2nd day after transplantation in all the patients. In 3 cases, it decreased after the initial peak and increased gradually to the normal level again within 2–4 weeks although, in the other 2 cases, it remained within the normal range after a slight decline. Serum 25-OHD and 24,25-$(OH)_2D$ level remained low for 4–8 weeks. Hypophosphatemia and/or hyperparathyroidism, which still remained after transplantation, may possibly stimulate 1α-hydroxylase in the transplanted kidney.

References

1   Seino Y, Tanaka H, Yamaoka K, Yabuuchi H: Circulating 1α,25-dihydroxyvitamin D_3 or 1α-hydroxyvitamin D_3 in normal men. Bone Mineral 1987;2:479–485.
2   Yamaoka K, Tanaka H, Kurose H, Shima M, Ozono K, Nakajima S, Seino Y: Effect of single oral phosphate loading on vitamin D metabolites in normal subjects and X-linked hypophosphatemic rickets. Bone Mineral 1989;7:159–169.
3   Lund B, Clausen E, Friedberg M, Moszkowicz M, Nielsen SP, Sorensen OH: Serum 1,25-dihydroxy-cholecalciferol in anephric, haemodialyzed and kidney-transplanted patients. Nephron 1980;25:30–33.
4   Shakhaee K, Brinker K, Helderman JH, Bengfort JL, Nicar MJ, Hull AR, Pak CYC: Disturbances in mineral metabolism after successful renal transplantation. Mineral Electrolyte Metab 1985;11:91–94.
5   Riancho JA, de Francisco ALM, del Arco C, Amado JA, Cotorruelo JG, Arias M, Gonzalez-Macias J: Serum levels of 1,25-dihydroxyvitamin D after renal transplantation. Mineral Electrolyte Metab 1988;14:332–337.
6   Garebedian M, Silive C, Levy-Bentolila D, Bourdeau A, Ulmann A, Nguyen TM, Lieberherr M, Broyer M, Balsan S: Changes in plasma 1,25- and 24,25-dihydroxyvitamin D after renal transplantation in children. Kidney Int 1981;20:403–410.
7   Schwartz GH, David DS, Riggio RR, Saville PD, Whitsell JC, Stenzel KH, Rubin AL: Hypercalcemia after renal transplantation. Am J Med 1970;49:42–51.
8   Alfrey AC, Jenkins D, Groth CG, Schorr WS, Gecelter L, Ogden DA: Resolution of hyperparathyroidism, renal osteodystrophy and metastatic calcification after renal homotransplantation. N Eng J Med 1968;279:1349–1356.

Dr. Shigeo Nakajima, Department of Pediatrics, Osaka University Medical School, 1-1-50 Fukushima, Fukushima-ku, Osaka 553 (Japan)

Morii H (ed): Calcium-Regulating Hormones. I. Role in Disease and Aging.
Contrib Nephrol. Basel, Karger, 1991, vol 90, pp 147–154

Circulating Bone Gla Protein in End-Stage Renal Disease Determined by Newly Developed Two-Site Immunoradiometric Assay

Kiyoshi Nakatsuka[a], *Takami Miki*[a], *Yoshiki Nishizawa*[a],
Tsutomu Tabata[b], *Takashi Inoue*[b], *Hirotoshi Morii*[a],
Etsuro Ogata[c,1]

[a]Second Department of Internal Medicine, Osaka City University Medical School,
Osaka, Japan; [b]Department of Internal Medicine, Inoue Hospital, Osaka, Japan;
[c]Fourth Department of Internal Medicine, University of Tokyo,
Faculty of Medicine, Tokyo, Japan

Several biochemical markers for bone metabolism have been postulated and available for use in the clinical research and management for metabolic bone diseases. Among these markers, bone Gla protein (BGP), also called osteocalcin, is an abundant noncollagenous protein found in bone matrix [1, 2]. This 49-residue peptide is synthesized by osteoblasts [3], which contains 3 residues of the vitamin K-dependent γ-carboxyglutamic acid (Gla) [4], remains in bone, and is released into the circulation [5].

The circulating nanomolar levels of BGP can be detected by radioimmunoassay (RIA) [6, 7], which is considered to reflect osteoblastic activity and is useful to assess bone remodeling, and, more specifically, bone formation [7, 8]. Bone histomorphometric studies have revealed that the serum BGP levels correlate with the parameters associated with bone formation rather than bone resorption in osteoporotic patients [9] and those with chronic renal failure [10].

However, multiple immunoreactive forms and/or fragments were demonstrated in the sera of patients with end-stage renal disease (ESRD), which are probably attributed to increased bone resorption and recognized by RIA [11]. Therefore, a methodology that specifically recognizes intact molecules of BGP has been desirable to clearly indicate the state of bone turnover in these patients. An immunoradiometric assay (IRMA) for

[1] We are grateful to Mr. M. Okada, Mitsubishi Petrochemical Co. Ltd., for his helpful technical assistance.

human BGP molecules has recently been developed, which excludes fragment forms of BGP found in the circulation. We validated this method and assessed its clinical availability, particularly in ESRD.

Subjects and Methods

BGP concentrations were measured in the sera of 55 patients with abnormal calcium metabolism, and 22 normal individuals by IRMA, which was developed by Mitsubishi Petrochemical Co. Ltd., Ibaraki, Japan. The serum samples obtained from these subjects were stored frozen at $-40\,°C$ until assayed. This IRMA system utilizes two types of monoclonal antibodies, which are specifically developed against c-terminal and midregion of human BGP (1–49) molecules, respectively. Standards of synthesized human BGP (1–49), human BGP (1-49, ^{17,21,24}Glu, Glu; glutamic acid) molecules, its fragments and serum samples were assayed according to the following procedure in duplicate:

Twenty five microliters of standard BGP or serum sample and 200 µl of borate buffer (pH 8.2, 0.1 M boric acid, 0.15 M NaCl, 0.02% Na_3N, 0.05% Tween 20, 0.1% bovine serum albumin) containing ^{125}I-labeled anti-BGP (34-49)Ab were pipetted into an assay tube. A polystyrene bead coated with anti-BGP (12-33)Ab was then put into the tube, which was incubated for 3 h at room temperature, shaking at 200 rpm. After aspirating the mixture, the bead was washed with 2 ml of distilled water. Radioactivity of the bead was counted for 1 min in a gamma spectrometer.

The patients were summarized as follows: (a) 10 women with postmenopausal osteoporosis without medications affecting calcium metabolism; (b) 3 men and 5 women with hypoparathyroidism who had discontinued active vitamin D supplements for about 4 weeks; (c) 3 men and 8 women with primary hyperparathyroidism before surgery; (d) 10 men and 10 women with ESRD undergoing dialysis; (e) 4 men and 2 women with hypercalcemia associated with malignancy. Serum BGP levels were determined using a standard curve of BGP (1-49^{17,21,24}Gla). In the case of high levels of BGP, BGP concentrations were calculated from the results of the samples diluted below 50 ng/ml.

Serum BGP concentrations in normal subjects were also determined by radioimmunoassay (RIA) with a RIA kit (CIS; Compagnie Oris Industrie Societe Anonyme, Saclay, France) to compare with those determined by IRMA.

In addition, serum BGP concentrations were determined by both the IRMA system and RIA with a RIA kit (Yamasa Co. Ltd., Chiba, Japan) in another group of 37 patients with ESRD, whose serum BGP levels were all below 50 ng/ml. Serum intact PTH levels and tartrate-resistant acid phosphatase (TRACP) activities were simultaneously assayed using an IRMA system (Allégro, Nichols Institute, Calif.) and the Kind-King method, respectively.

Differences between groups were analyzed using the unpaired t test. Differences and correlations between the assays were assessed by the paired t test and linear regression analysis, respectively.

Results

The corrected dose response of radioactivity was observed linearly plotted over the range 1–64 ng/ml of human BGP (1–49,^{17,21,24}Gla), human

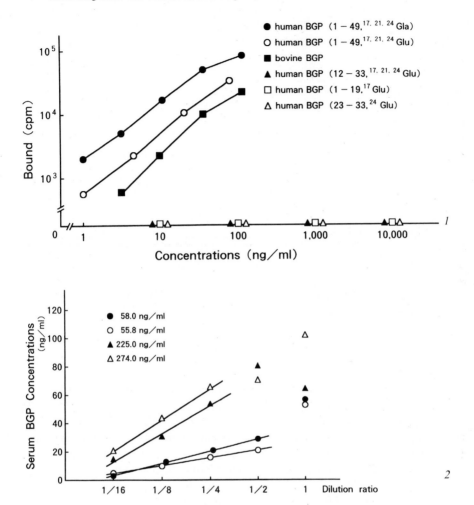

Fig. 1. Cross-reactivities of bone Gla protein molecules in IRMA.

Fig. 2. A series of dilutions of sera in dialysis patients. Note parallelism in the range below 50 ng/ml.

BGP (1–49,[17,21,24]Glu), and bovine BGP, while fragments of the molecules (1–19, 12–33, 23–33) were not recognized by this assay system (fig. 1).

A series of dilutions of sera in dialysis patients with high BGP levels demonstrated parallelism in the range below 50 ng/ml (fig. 2). The precision of the assay for sera with variable BGP levels was determined with the intra-assay variance, CV: 2.3–2.4%, and the reproducibility with the

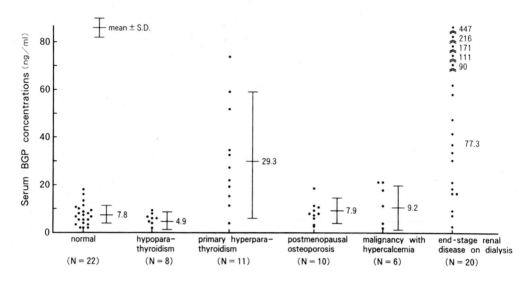

Fig. 3. Serum BGP concentrations in normal subjects and in patients with abnormal calcium metabolism.

inter-assay variance, CV: 2.2–5.2%. Furthermore, the analytical recovery ranged from 104 to 116%.

Serum BGP levels in healthy controls measured by IRMA were 7.8 ± 3.6 ng/ml (SD) (3.9–12.1); postmenopausal osteoporosis 7.9 ± 3.9 ng/ml (3.2–17.4); hypoparathyroidism 4.9 ± 2.9 ng/ml (0.2–10.1), this value being significantly lower than normal controls (p < 0.05); primary hyperparathyroidism 29.3 ± 20.6 ng/ml (0.3–43.8), the value being significantly higher than normal controls (p < 0.01); chronic renal failure receiving dialysis, ranging from 6.3 to 447 ng/ml; hypercalcemia associated with malignancy 9.2 ± 7.9 ng/ml (1.1–19.6) (fig. 3).

A close positive correlation was demonstrated between serum BGP concentrations determined by IRMA and those simultaneously measured by conventional RIA in normal individuals (r = 0.935, p < 0.001, n = 16; fig. 4) as well as in dialysis patients (r = 0.918, p < 0.001, n = 37; fig. 5). However, the levels of BGP determined by IRMA were estimated significantly lower than those by conventional RIA (23.6 ± 9.8 vs. 29.8 ± 9.1 ng/ml, p < 0.00001).

Positive correlations were additionally demonstrated between serum BGP levels and serum intact PTH levels (r = 0.462, p < 0.01, n = 32), and serum TRACP activities (r = 0.443, p < 0.01, n = 35) in dialysis patients with ESRD (fig. 6).

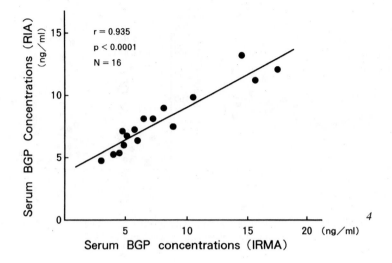

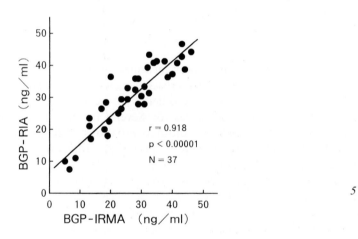

Fig. 4. Relationship in serum bone Gla protein levels between **IRMA** and **RIA** in normal individuals.

Fig. 5. Relationship in serum bone Gla protein levels between **IRMA** and **RIA** in end-stage renal disease.

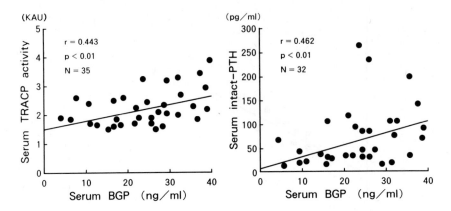

Fig. 6. Relationships between serum bone Gla protein levels and serum intact PTH levels, and serum tartrate-resistant acid phosphatase activities (TRACP) in end-stage renal disease.

Discussion

Circulating levels of BGP have been measured by RIA using antiserum developed against bovine BGP, which cross-reacts with human BGP [6, 7]. Most of these antisera have been considered to recognize the carboxy-terminal regions of the BGP molecules [6]. Several factors are known to affect serum BGP levels, such as diurnal variations [12], aging [13], stimulations of 1,25-dihydroxyvitamin D_3 [14], and various diseases with increased bone remodeling [7, 15, 16].

However, BGP is rapidly cleared from the circulation and mainly metabolized in the kidney [5] and its serum levels are elevated due to decreased renal clearance [17]. Thus, elevated serum levels of BGP are caused by both the delay of BGP metabolism by the kidney and increased bone formation in patients with ESRD receiving long-term dialysis. Moreover, multiple immunoreactive fragments of BGP are found in uremic sera, not in those of normal individuals [11]. Considering these circumstances, assays which specifically recognize intact molecules of BGP have been needed to estimate the bone formation, particularly in individuals with decreased renal functions.

On the other hand, in the recent decade, monoclonal antibodies specific to polypeptide, mainly to peptide hormones, have been applied to two-site IRMA for measurements of their circulating levels [18, 19]. This methodology has made it possible to detect intact molecules of the peptide hormones and been considered useful to assess endocrine functions [20]. We presented an IRMA system extended to the measurement of circulating intact BGP levels, resulting in both a good sensitivity and a good specificity, and determined serum BGP levels in patients with several types of abnormal calcium metabolism using this assay.

Attention, however, should be paid to determining the BGP levels of sera in such patients with ESRD on dialysis, which may be underestimated because of a relative shortage of monoclonal antibodies in BGP-rich sera. This problem, known as the hook phenomenon, is characteristic in two-site IRMA and could be avoided by diluting samples to levels below 50 ng/ml.

Further investigations should be performed to ensure that serum intact BGP levels determined by this assay mainly reflect BGP production in bone, particularly in renal insufficiency. We concluded that the above-mentioned BGP-IRMA system, with an easy and rapid procedure, will not only provide information to estimate bone turnover but also contribute to clarifying the metabolism of BGP itself.

Summary

As a marker for bone formation, bone Gla protein (BGP) levels in the circulation have been measured in clinical research and management for metabolic bone diseases. We evaluated the clinical availability of a newly developed two-site immunoradiometric assay (IRMA) for human BGP and determined the serum BGP concentrations using this methodology in patients with abnormal calcium metabolism including those with end-stage renal disease undergoing maintenance dialysis. A cross-reactivity test revealed that this assay system specifically recognizes intact molecules (1–49) of BGP and excludes fragments of the molecules (1–19, 12–33, 23–33). Serum BGP levels in dialysis patients were positively correlated with those by conventional radioimmunoassay (RIA) ($r = 0.918$, $p < 0.00001$, $n = 37$) as well as normal individuals ($r = 0.935$, $p < 0.0001$, $n = 16$). However, the levels of BGP determined by IRMA were estimated to be significantly lower than those by RIA (23.6 ± 9.8 vs. 29.6 ± 9.1 ng/ml, $p < 0.00001$).

These results suggest that this IRMA system, with a rapid and easy procedure, excludes fragment forms of BGP in the circulation, which are found in uremic sera and probably attributed to increased bone resorption. Further studies are needed to ensure that serum intact BGP levels mainly reflect BGP production in osteoblasts, particularly in end-stage renal disease.

References

1   Hauschka PV, Lian JB, Gallop PM: Direct identification of the calcium-binding amino acid, carboxyglutamate, in mineralized tissue. Proc Natl Acad Sci USA 1975;72:3925.
2   Price PA, Otsuka AS, Poser JW, Kristaponis I, Raman N: Characterization of a

carboxyglutamic acid-containing protein from bone. Proc Natl Acad Sci USA 1976;73:1447.

3   Lian JB, Friedman PA: The vitamin K-dependent synthesis of gamma carboxyglutamic acid by bone microsome. J Biol Chem 1978;253:6623.

4   Poser JW, Esch FS, Ling NC, Price PA: Isolation and sequence of the vitamin K-dependent protein from human bone. J Biol Chem 1980;255:8685.

5   Price PA, Williamson MK, Lothringer JW: Origin of the Vitamin K-dependent bone protein found in plasma and its clearance by kidney and bone. J Biol Chem 1981;256:12760.

6   Price PA, Nishimoto SK: Radioimmunoassay for the vitamin K-dependent protein of bone and its discovery in plasma. Proc Natl Acad Sci USA 1980;77:2234.

7   Price PA, Parthemore JG, Deftos LJ: New biochemical marker for bone metabolism: Measurement by radioimmunoassay of bone-Gla protein in the plasma of normal subjects and patients with bone disease. J Clin Invest 1980;66:878.

8   Slovic DM, Gundberg CM, Neer RM, Lian JB: Clinical evaluation of bone turnover by serum osteocalcin measurements in a hospital setting. J Clin Endocrinol Metab 1984;59:228.

9   Brown JP, Delmas PD, Malaval L, Edouard C, Chapuy MC, Meunier PJ: Serum bone Gla-protein: A specific marker for bone formation in postmenopausal osteoporosis. Lancet 1984;i:1091.

10 Malluche HH, Faugere MC, Fanti P, Price PA: Plasma levels of bone Gla-protein reflect bone formation in patients on chronic maintenance dialysis. Kidney Int 1984;36:869.

11 Gundberg CM, Weinstein RS: Multiple immunoreactive forms of osteocalcin in uremic serum. J Clin Invest 1986;77:1762.

12 Gundberg CM, Markowitz ME, Mizruchi M, Rosen JF: Osteocalcin in human serum: A circadian rhythm. J Clin Endocrinol Metab 1985;60:736.

13 Delmas PD, Stenner D, Wahner HW, Mann KG, Riggs BL: Increase in serum bone γ-carboxyglutamic acid protein with aging in women. J Clin Invest 1983;71:1316.

14 Morkowitz ME, Gundberg CM, Rosen JF: The circadian rhythm of serum osteocalcin concentrations: Effect of 1,25-dihydroxy-vitamin D administration. Calcif Tissue Int 1987;40:179.

15 Lukert BP, Higgins JC, Stodkopf MM: Serum osteocalcin is increased in patients receiving glucocorticoids. J Clin Endocrinol Metab 1986;62:1056.

16 Delmas PD, Demiaux B, Malaval L, Chapuy MC, Edouard C, Meunier PJ: Serum gamma carboxyglutamic acid-protein in primary hyperparathyroidism and in malignant hypercalcemia. J Clin Invest 1986;77:985.

17 Delmas PD, Wilson DM, Mann KG, Riggs BL: Effect of renal function on plasma levels of bone Gla-protein. J Clin Endocrinol Metab 1983;57:1028.

18 Hunter WM, Bennie JG, Brock DJM, Heyningen VV: Monoclonal antibodies for use in an immunoradiometric assay for α-fetoprotein. J Immunol Methods 1982;50:41.

19 Belanger A, Cote J, Lavoie M: The production of high affinity monoclonal antibodies to human chorionic gonadotropin and their application to immunoradiometric assay. J Immunoassay 1985;7:37.

20 Nussbaum SR, Zahradnik RJ, Lavigne JR, Brennan GL, Nozawa-Ung K, Kim LV, Keutmann HT: Highly sensitive two-site immunoradiometric assay of parathyrin, and its clinical utility in evaluating patients with hypercalcemia. Clin Chem 1987;33:1364.

Kiyoshi Nakatsuka, MD, Second Department of Internal Medicine,
Osaka City University Medical School, 1-5-7 Asahi-machi,
Abeno-ku, Osaka 545 (Japan)

Treatment of Calcium Abnormalities in Chronic Renal Diseases

Morii H (ed): Calcium-Regulating Hormones. I. Role in Disease and Aging.
Contrib Nephrol. Basel, Karger, 1991, vol 90, pp 155–160

Role of Calcium Supplementation in Patients with Mild Renal Failure

Kunitoshi Iseki, Shinichiro Osato, Kaoru Onoyama, Masatoshi Fujishima[1]

Second Department of Internal Medicine, Kyushu University, Fukuoka, Japan

Prevention of the progression of renal failure is an ultimate goal for the therapy of patients with renal disease. There is no known unequivocal determinant of the progression of renal failure [1]. A low-protein diet was shown to be effective in the prevention of renal failure [2, 3]. However, the roles of protein and phosphate intake have not been definitely separated. Dietary phosphate restriction is effective to delay the occurrence of secondary hyperparathyroidism [4]. Renal calcium deposition will be reduced by phosphate restrictions [5]. In the rat study, phosphate restriction per se is shown to be effective to prevent the progression of renal failure [6]. However, the clinical application of phosphate restriction is yet to be known.

Calcium carbonate is effective as a phosphate binder [7] and a hypotensive agent [8]. Calcium, phosphate and PTH are mutually responsible for maintenance of blood pressure in renal failure patients [9], and the blood pressure response to oral calcium administration is exaggerated in the presence of high PTH [10]. On the other hand, excess calcium may cause hypercalcemia and will further deteriorate renal function [11]. It is still unknown whether calcium carbonate supplementation is feasible in patients with early renal failure and is effective to prevent the progression of renal failure [12].

We tested a hypothesis that calcium carbonate is useful for early renal failure patient through its phosphate binding and hypotensive action.

[1] We are grateful for the help of Dr. S. Fujii, Dr. K. Ohtsubo, and Dr. K. Tsukiji.

Materials and Methods

To investigate the effect of calcium carbonate supplementation in patients with renal disease, we have planned a prospective, randomized, double-blind study. Thirty-six patients, 16 females and 20 males, were recruited from our outpatient renal clinic. Patients with renal disease who has renal biopsy or renal dysfunction with high serum creatinine (> 1.4 mg/dl), and/or persistent proteinuria microhematuria were selected for the study. Hypertensive patients and/or those on antihypertensive medications were excluded from the study. Their mean age was 48.9, ranging from 27 to 81, and serum creatinine was 2.1 mg/dl, ranging from 0.8 to 4.9 mg/dl. Original kidney diseases were IgA nephropathy [11], non IgA glomeru-lonephritis [12], chronic interstitial nephritis [6], chronic renal failure [3], rheumatoid arthritis [2], periarteritis nodosa [1], and polycystic kidney disease [1].

Patients were visiting every 4 weeks. After a 2-month control period (V1, V2), they were treated with drugs, either calcium carbonate or placebo, three times a day, 4 capsules each, for 3 months (V3, V4, V5). The daily dosage of calcium was 6 g. The number of capsules left in the bottle was checked at each visit. Drugs were prepared, randomized, and kept by our pharmacists. No change in dietary prescription was made during the study. They were asked to bring 24-hour urine at each visit. Some used Urinmate and brought 1/50 vol of the total urine [13].

Blood and urine chemistry were measured at every visit. Plasma ionized calcium concentration was measured immediately after blood drawing. PTH, C-terminal, and plasma catecholamines were measured at second (V2) and fourth (V4) visits. Blood pressure and heart rate were measured by automatic sphygmomanometer at supine and standing position. Mean blood pressure was calculated as diastolic blood pressure plus one-third of the pulse pressure.

They were explained the protocol and consented to participate in the study. Data were expressed as mean ±SEM.

Results

There was no significant change in serum calcium, phosphate, BUN, serum creatinine, total protein concentration, and plasma ionized calcium concentration both in placebo group (n = 17) and in calcium carbonate group (n = 19) throughout the study period. At V2, plasma ionized calcium concentration (Ca ion) was 1.06 ± 0.02 mmol in the placebo group and 1.06 ± 0.02 mmol in the calcium carbonate group. At V4, it was 1.06 ± 0.02 mmol in the former and 1.09 ± 0.02 mmol in the latter. In the calcium carbonate group, 2 cases of hypercalcemia (> 10.5 mg/dl) were noted. Nephrocalcinosis became worse in one of the cases who also had an episode of macrohematuria. One patient in the calcium carbonate group had hemorrhoidal bleeding. C-terminal PTH, plasma epinephrine and norepinephrine concentration were not different between the groups either at V2 or at V4.

There was no significant change is systolic, diastolic, or mean blood pressure either in placebo or in calcium carbonate group throughout the study period. At V2, supine mean blood pressure was 102.6 ± 3.3 mm Hg

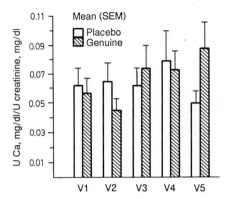

Fig. 1. Serial change in urinary excretion of calcium (mg/dl) corrected by urinary creatinine (mg/dl). There was no significant difference between the groups.

in the placebo group and 104.1 ± 3.3 mm Hg in the calcium carbonate group, respectively. At V4, it was 101.7 ± 3.3 and 105.6 ± 3.0 mm Hg, respectively. There were no postural changes in blood pressure in both groups. However, a significant drop in systolic blood pressure in the standing position was noted after 1 month of calcium carbonate (V3), otherwise there was no significant change in blood pressure in the calcium carbonate group.

There was no significant change in 24-hour creatinine clearance in both groups throughout the study period. In the placebo group, it was 61.1 ± 8.4 ml/min/1.73 m^2 at V2 and 61.1 ± 10.4 ml/min/1.73 m^2 at V4. In the calcium carbonate group, it was 43.9 ± 5.1 ml/min/1.73 m^2 at V2, and 40.0 ± 4.3 ml/min/1.73 m^2 at V4. Urine urea-N (mg/dl) corrected by urine creatinine (mg/dl) was constant throughout the study period in both groups. At V1, it was 7.4 ± 0.7 in the placebo group and 7.4 ± 0.8 in the calcium carbonate group. Therefore, protein intake during the study period seemed constant in both groups. Urine calcium (mg/dl) corrected by urine creatinine (mg/dl) was not changed during the study and was not different between the groups (fig. 1). Urinary phosphate (mg/dl) corrected by urine creatinine (mg/dl) was significantly decreased in the calcium carbonate group compared to that in the placebo group at V3, V4, and V5, respectively ($p < 0.01$, Anova: fig. 2). There was a significant correlation between the urinary excretion of phosphate (mg/day) and that of urea-N (g/day) in both groups. The two lines were not parallel, but the y-intercept of the two lines was significantly different ($p < 0.01$, fig. 3).

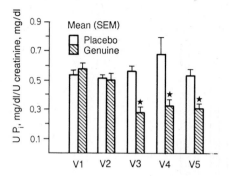

Fig. 2. Serial changes in urinary phosphate excretion (mg/dl), corrected by urinary creatinine (mg/dl). At V3, V4, and V5, there was a significant decrease in urinary phosphate excretion in the calcium carbonate group (ANOVA, *p < 0.01).

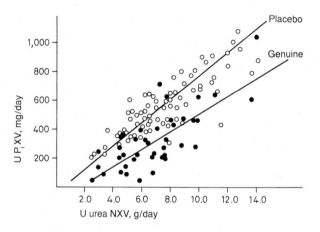

Fig. 3. There was a significant correlation between urinary phosphate excretion (mg/ day) and urinary excretion of urea-N (g/day) in both groups. The two lines were not parallel, although the y-intercept was significantly different (p < 0.01).

Discussion

The results of the present study demonstrate that therapy with 6 g/day of calcium carbonate is feasible in early renal failure patients. Urinary phosphate excretion was reduced to 50% of that in the placebo group. Dietary phosphate restriction per se also delays secondary hyperparathyroidism [4]. Introduction of a low-protein diet may affect the quality of life [2]. Accordingly, long-term administration of calcium carbonate would be one of the choices to prevent the occurrence of secondary hyperparathyroidism.

A previous study has shown that phosphate restrictions per se has a beneficial effect in the prevention of progression of renal failure in the remnant kidney model in the rat [6]. Conversely, a high phosphate intake has been shown to accelerate renal function in rats with reduced renal mass [14]. Although others denied the role of phosphate depletion [15], Tomford et al. [16] reported the significance of thyroid hormone, not PTH, in the prevention of progression of renal failure.

In humans, the role of phosphate and PTH in progression of renal failure is contradictory [1]. There was a significant correlation between serum phosphate concentration and the rate of progression of renal failure [17], but others [1, 18] did not confirm this. Massry et al. [19] reported that higher PTH is associated with rapid progression of renal failure, but the others reported the opposite result [1]. Fournier et al. [12] have shown that 3 g/day of calcium carbonate in association with nonhypercalcemic doses of $25(OH)D_3$ has decreased the slope of $1/Pcr$ in early renal failure, although the clinical application is yet to be shown.

The hypotensive effect was not evident in the present study. Concerning hypertension secondary to renal disease, there are contradictory reports regarding the level of blood pressure and the progression of renal failure [1].

These data and the results of the present study, therefore, indicate that long-term administration of calcium carbonate should be performed to see if reduced phosphate intake per se has a beneficial effect on the progression of renal failure. Frequent and meticulous follow-up would be necessary to prevent complications of a possible hypercalcemia.

Summary

A low-protein diet has been shown to be effective in the prevention of progression of renal failure. However, the role of protein and that of phosphate intake have not been definitely separated. To determine the use of calcium ($CaCO_3$) as a phosphate binder and a possible hypotensive agent, we performed a prospective, double-blind study of oral calcium in patients with mild-to-moderate renal failure (SCr < 5.0 mg/dl). Thirty-six patients (20 m, 16 f) were studied. After a 2-month control period, either $CaCO_3$ (6 g/day, 12 Cap) or placebo was given for 3 months. There was no significant difference in BUN, S-Sr, S-Ca, S-Pi, 24-hour Ccr, and blood pressure during the study period both in the genuine and the placebo group. Urine phosphate was decreased by about 50% in the $CaCO_3$ group. Calcium carbonate is feasible as a phosphate binder for patients with mild-to-moderate renal failure.

References

1   Walser M: Progression of renal failure in man (Editorial review). Kidney Int 1990; 22:1148–1152.

2   Ihle BU, Becker GJ, Whitworth JA, Charlwood RA, Kincaid-Smith PS: The effect of
    protein restriction on the progression of renal insufficiency. N Engl J Med 1989;
    321:1773–1777.
3   Rosman JB, terWee PM, Meijer S, Piers-Becht TP, Sluiter WJ, Donker AJ: Prospective
    randomized trial of early dietary protein restriction in chronic renal failure. Lancet
    1984;ii:1291–1296.
4   Llach F, Massry SG: On the mechanism of secondary hyperparathyroidism in moderate
    renal insufficiency. J Clin Endocrinol Metab 1985;61:601–606.
5   Ibels LS, Alfrey AC, Haut L, Huffer WE: Preservation of function in experimental renal
    disease by dietary phosphate restriction. N Engl J Med 1978;298:122–126.
6   Lumlertgul D, Burke TJ, Gillum DM, Alfrey AC, Harris DC, Hammond WS, Schrier
    RW: Phosphate depletion arrests progression of chronic renal failure independent of
    protein intake. Kidney Int 1986;29:658–666.
7   Slatopolsky E, Weerts C, Lopez-Hilker S, Norwood K, Zink M, Windus D, Delmez J:
    Calcium carbonate as a phosphate binder in a patient with chronic renal failure
    undergoing dialysis. N Engl J Med 1986;315:157–161.
8   McCarron DA, Morris CD: Blood pressure response to oral calcium in persons with
    mild to moderate hypertension. A randomized, double-blind, placebo-controlled,
    crossover trial. Ann Intern Med 1985;103:825–831.
9   Massry SG, Iseki K, Campese VM: Serum calcium, parathyroid hormone, and blood
    pressure. Am J Nephrol 1986;6(suppl 1):S19–S28.
10 Lyle RM, Melby CL, Hyner GC: Metabolic differences between subjects whose blood
    pressure did or did not respond to oral calcium supplementation. Am J Clin Nutr
    1988;47:1030–1035.
11 Massry SG, Kaptein EM: Hypercalcemia and hypocalcemia; in: Textbook of Nephrol-
    ogy. Baltimore, Williams & Wilkins, 1989, ed 2, pp 300–311.
12 Fournier A, Moriniere P, Renaud H: The approach to the treatment of secondary
    hyperparathyroidism in early renal failure. Am J Nephrol 1988;8:170–172.
13 Tochikubo O, Uneda S, Kaneko Y: Single portable device for sampling a whole day's
    urine and its application to hypertensive patients. Hypertension 1983;5:270–274.
14 Haut LL, Alfrey AC, Guggenheim S, Buddington B, Schrier RW: Renal toxicity of
    phosphate in rats. Kindey Int 1980;17:722–731.
15 Laouari D, Kleinknecht C, Gubler MC, Broyer M: Adverse effect of proteins on
    remnant kidney: Dissociation from that of other nutrients. Kidney Int 1983;24(suppl
    6):S248–S253.
16 Tomford RC, Karlingsky ML, Buddington B, Alfrey AC: Effect of thyroparathyroidec-
    tomy and parathyroidectomy on renal function and the nephrotic syndrome in rat
    nephrotoxic serum nephritis. J Clin Invest 1981;68:655–664.
17 Barsotti G, Giannoni A, Morelli E, Lazzeri M, Vlamis I, Baldi R, Giovannetti S: The
    decline of renal function slowed by a very low phosphate intake in chronic renal patients
    following a low nitrogen diet. Clin Nephrol 1984;21:54–59.
18 Gretz N, Meisinger E, Strauch M: Correlation between serum phosphate concentration
    and rate of progression of chronic renal failure. Proc EDTA-ERA 1985;22:1148–1152.
19 Massry SG, Frohling PT, Akmal M: Is parathyroid hormone involved in the progres-
    sion of renal failure? In: Prevention of Progressive Uremia. New York, Field & Wood,
    1989, pp 125–127.

Dr. K. Iseki, 3rd Department of Internal Medicine, Ryukyu University,
207 Uehara, Nishihara, Okinawa 903-01 (Japan)

Morii H (ed): Calcium-Regulating Hormones. I. Role in Disease and Aging.
Contrib Nephrol. Basel, Karger, 1991, vol 90, pp 161–165

Long-Term Effects of Calcium Antagonists and Angiotensin-Converting Enzyme Inhibitors in Patients with Chronic Renal Failure of IgA Nephropathy

Mikio Okamura, Yoshiharu Kanayama, Nobuo Negoro,
Takatoshi Inoue, Tadanao Takeda

First Department of Internal Medicine, Osaka City University Medical School,
Osaka, Japan

Various hemodynamic and dietary factors have been shown to alter the rate of decline in renal function [1]. Recent studies have shown that dietary protein restriction and normalization of glomerular hypertension with angiotensin-converting enzyme (ACE) inhibitor treatment slow the rate of this decline in renal function [2, 3]. However, there is limited information available regarding the long-term effects of the calcium (Ca) antagonist in human renal diseases [4]. The purpose of this study is to evaluate the long-term effect of the Ca antagonist and ACE inhibitor on renal function in hypertensive patients with chronic renal failure of IgA nephropathy.

Patients and Methods

A total 20 patients with chronic renal failure of IgA nephropathy entered the study. A diagnosis of IgA nephropathy was made by renal biopsy. All were hypertensive, blood pressure was more than 160 mm Hg systolic and 95 mm Hg diastolic. Chronic renal failure was defined by serum creatinine levels over 1.5 mg/dl. Nine patients received Ca antagonists (nifedipine 6, nicardipine 3) and 11 received ACE inhibitors (captopril 4, enalapril 7). The patients characteristics of Ca antagonist and ACE inhibitor treatment are given in table 1. Mean pretreatment blood pressure was $183 \pm 22/107 \pm 13$ mm Hg in the Ca antagonist-treated patients and $173 \pm 15/107 \pm 11$ mm Hg in the ACE inhibitor-treated patients. Other characteristics including age, duration of IgA nephropathy and renal function measured by blood urea nitrogen (BUN), serum creatinine and endogenous creatinine clearance, were also similar in both treatment groups. 24-hour proteinuria was almost the same in both groups. The effects of treatment on renal function was determined by measuring BUN and serum creatinine monthly. Endogenous creatinine clearance and 24 hour proteinuria was measured at baseline and at 3, 6 and 12 months after the initiation of treatment.

Table 1. Patient characteristics in Ca antagonist or ACE inhibitor treatment

	Ca antagonist (nifedipine 6, nicardipine 3)	ACE inhibitor (captopril 4, enalapril 7)
Number of patients	9	11
Sex M:F	4:5	7:4
Age, years	48.3 ± 6.4	53.6 ± 10.0
Duration, years	11.7 ± 4.2	10.1 ± 3.2
Systolic blood pressure (SBP), mm Hg	183 ± 22	173 ± 15
Diastolic blood pressure (DBP), mm Hg	107 ± 13	107 ± 11
Blood urea nitrogen (BUN), mg/dl	25.3 ± 9.1	22.9 ± 4.1
Creatinine, mg/dl	1.9 ± 0.7	1.8 ± 0.2
Creatinine clearance (C_{cr}), ml/min	44 ± 12	41 ± 11
24 hour urinary protein, g/day	2.0 ± 1.1	2.0 ± 1.2

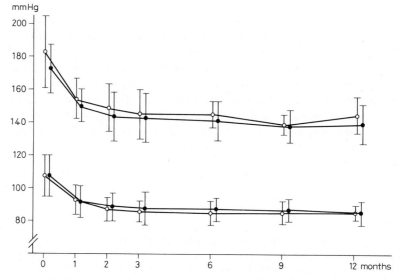

Fig. 1. Blood pressure response in patients with chronic renal failure of IgA nephropathy. ○ = Ca antagonist treatment; ● = ACE inhibitor treatment.

Results

Blood Pressure

Blood pressure response for Ca antagonist or ACE inhibitor treatment is shown in figure 1. In both treatment groups systolic blood pressure (SBP) and diastolic blood pressure (DBP) were reduced at 1 month of

Table 2. Long-term effects of Ca antagonist and ACE inhibitor on blood pressure and renal function in patients with chronic renal failure of IgA nephropathy

	Month 0	Month 6	Month 12
Ca antagonist (n = 9)			
SBP	183 ± 22	145 ± 8	145 ± 11
DBP	107 ± 13	86 ± 7	87 ± 4
BUN	25.3 ± 9.1	22.9 ± 5.4	25.1 ± 6.1
Creatinine	1.9 ± 0.7	1.9 ± 0.5	2.0 ± 0.6
C_{cr}	44 ± 12	40 ± 9	42 ± 13
24 hour urinary protein	2.0 ± 1.1	1.9 ± 0.9	2.0 ± 1.0
ACE inhibitor (n = 11)			
SBP	173 ± 15	141 ± 12	139 ± 12
DBP	107 ± 13	89 ± 7	87 ± 7
BUN	22.9 ± 4.1	22.3 ± 4.0	22.1 ± 4.5
Creatinine	1.8 ± 0.2	1.9 ± 0.5	2.0 ± 0.6
C_{cr}	41 ± 11	44 ± 10	43 ± 11
24 hour urinary protein	2.0 ± 1.2	1.5 ± 1.1	1.2 ± 1.1*

*$p < 0.05$. For abbreviations and units see table 1.

treatment. These reductions in SBP and DBP were statistically significant ($p < 0.01$) compared to initial levels in both treatment groups. Thereafter, similar blood pressure levels were maintained in most patients until the end of the study.

Renal Function

The long-term effects of treatment on renal function are shown in table 2. Mean BUN and serum creatinine did not change. In Ca antagonist treatment 5 patients had a small increase and 4 patients had a decrease in serum creatinine, while in ACE inhibitor treatment 6 patients had a small increase and 6 patients had a decrease in serum creatinine. Mean endogenous creatinine clearance levels were unchanged during the study in Ca antogonist and ACE inhibitor treatment. At the end of the study, renal function in Ca antagonist treatment was not statistically different to that in ACE inhibitor treatment.

Proteinuria

The changes in mean 24-hour proteinuria during the study are contrasting in both treatment groups. In Ca antagonist treatment, 24-hour proteinuria was not changed compared to initial levels during the study, while proteinuria was decreased significantly compared to initial levels

$(2.0 \pm 1.1$ to 1.2 ± 1.1 g, $p < 0.05$) during ACE inhibitor treatment, thus result-ing in a total fall of 40%. This reduction of proteinuria was also statistical-ly significant between Ca antagonist treatment and ACE inhibitor treatment at the end of the study ($p < 0.05$).

Discussion

Systemic hypertension has been recognized as a risk factor for the progression of renal failure. Clinical studies have not uniformly demon-strated slowing of progressive renal disease with antihypertensive treat-ment. Recent studies have shown that normalization of glomerular hypertension with ACE inhibitors slow the rate of decline in renal function in experimental renal disease [3]. Although both ACE inhibitors and Ca antagonists reduce blood pressure by vasodilation, they have a distinct mechanism of action. ACE inhibitors have a vasodilator effect mainly on the efferent glomerular arterioles, while Ca antagonists act mainly on the afferent arterioles. Thus, Ca antagonists would probably increase glomeru-lar capillary pressure. On the other hand, it is reported that acute infusion of diltiazem or verapamil reduces glomerular capillary pressure in rats undergoing subtotal nephrectomy [5]. In addition, diltiazem reduces platelet aggregation [6] and this may slow the rate of progression of renal disease. Because of these countervailing possibilities, it is difficult to predict what effect Ca antagonist might have on the progression of chronic renal failure [4].

Eliahou et al [7] investigated the effect of the Ca antagonist nisoldipine on the progression of renal disease in patients with chronic renal failure of different etiologies [7]. The rate of progression of renal failure was signifi-cantly less in the nisoldipine-treated group than in combination of diuret-ics, beta-blockers and vasodilators. In patients with diabetic nephropathy treatment of captopril or nifedipine caused a contrasting effect [8]. Capto-pril treatment was associated with a 40% decrease in urinary albumin excretion, while nifedipine induced a rise in urinary albumin excretion of 40%.

In our study, Ca antagonist and ACE inhibitor were equally effective in reducing blood pressure in patients with hypertensive patients with IgA nephropathy. Renal functions determined by BUN, serum creatinine and endogenous creatinine clearance were not changed significantly in both treatment groups for a period of 12 months. The changes of urinary protein excretion were contrasting. Urinary protein was decreased in ACE inhibitor treatment, while it was unchanged compared to the initial level in Ca antagonist treatment. From these results, the long-term treatment of Ca antagonist and ACE inhibitor are equally effective and preserved renal function in hypertensive patients with chronic renal failure of IgA nephropathy.

As initiation and progression of chronic renal failure is caused by diverse etiology, controlled clinical trials are needed before these treatment can be recommended clinically for the presevation of renal function.

Summary

Long-term effects of Ca antagonist and ACE inhibitor on renal function in hypertensive patients with chronic renal failure of IgA nephropathy were studied. Both Ca antagonists and ACE inhibitors were equally effective in reducing blood pressure.

References

1   Klahr S, Schreiner G, Ichikawa I: The progression of renal disease. N Engl J Med 1988;318:1657–1666.
2   Nath KA, Kren SM, Hostetter TH: Dietary protein restriction in established renal injury in rat. J Clin Invest 1986;78:1199–1205.
3   Anderson S, Rennke HG, Brenner BM: Therapeutic advantage of converting enzyme inhibitors in arresting progressive renal disease associated with systemic hypertension in rats. J Clin Invest 1986;77:1993–2000.
4   Krishna GG, Narins RG: Calcium channel blockers. Progression of renal disease. Circulation 1989;80(suppl IV):IV47–51.
5   Anderson S, Clarye LE, Riley SL, Troy JL: Acute infusion of calcium channel blockers reduces glomeruler capillary pressure in rats with reduced renal mass (abstract). Kidney Int 1987;33:370.
6   Mehta J, Mehta P, Ostrowski N: Calcium channel blocker diltiazem inhibits platelet activation and stimulates vascular prostacyclin synthesis. Am J Med Sci 1986;291:20–24.
7   Eliahou HE, Cohen D, Hellberg B, Ben-David A, Herzog D, Schecher P, Kapuler S, Kogan N: Effect of the calcium channel blocker nisoldipine on the progression of chronic renal failure in man. Am J Nephrol 1988;8:285–290.
8   Mimran A, Insula A, Ribstein J, Monnier L, Bringer J, Mirouze J: Contrasting effects of captopril and nifedipine in normotensive patients with incipient diabetic nephropathy. J Hypertens 1988;6:919–923.

Dr. Mikio Okamura, First Department of Internal Medicine, Osaka City University Medical School, 1-5-7, Asahi-cho, Abeno-ku, Osaka 545 (Japan)

Morii H (ed): Calcium-Regulating Hormones. I. Role in Disease and Aging.
Contrib Nephrol. Basel, Karger, 1991, vol 90, pp 166–182

Oral and Parenteral Calcitriol for the Management of End-Stage Renal Disease[1]

Jack W. Coburn, Isidro B. Salusky, Keith C. Norris, William G. Goodman

Medical and Research Services, West Los Angeles Veterans Affairs Medical Center (Wadsworth Division) and Sepulveda Veterans Affairs Medical Center and the Departments of Medicine and Pediatrics, UCLA School of Medicine, Los Angeles, Calif., USA

Several discoveries have contributed to improved management of the bone disease arising due to secondary hyperparathyroidism in patients with end-stage renal disease: It was found that the kidney is the major organ responsible for generating calcitriol, the active, hormonal form of vitamin D [1]. Subsequently, calcitriol and its analogue, 1-α-hydroxyvitamin D, were synthesized and then used to treat patients with renal failure [2–4], with more than 10 years' experience with their use. Other data indicated that the absolute or relative deficiency of calcitriol is a major factor leading to increased secretion of parathyroid hormone (PTH) in patients with renal failure [5, 6]. Finally, parenteral calcitriol was introduced, and data suggest that giving the sterol in this form is highly effective in suppressing serum iPTH levels in patients with end-stage renal disease [7]. What are effects of calcitriol in patients with renal failure and how do the effects of oral calcitriol compare to the parenteral form? Since parenteral calcitriol is primarily applicable to patients undergoing hemodialysis because of the access to their circulation, comparisons will be limited to data from patients undergoing hemodialysis.

Effects of Calcitriol in Uremic Patients

The actions of calcitriol in patients with end-stage renal disease are summarized in table 1, and these effects are highly relevant to its use in renal osteodystrophy. Most of these effects have been evaluated following

[1] Some of the work cited was supported by research funds from the Department of Veterans Affairs.

Table 1. Actions of calcitriol in patients with renal failure: benefits/side effects

Action	Potential benefit	Potential side-effects
Increase Ca absorption	correct hypocalcemia	limits dosage
Increase PO₄ absorption	rare (low PO₄)	raises serum PO₄
Increase Mg absorption	no benefit	hypermagnesemia
Increase serum Al?	? trivial effect	worsen Al loading (?)
Increase zinc absorption	correct Zn deficiency	?raise plasma zinc
Mucosal hypertrophy	uncertain	uncertain
Raise serum calcium	correct hypocalcemia	worsen hypercalcemia
Suppress iPTH levels		
direct parathyroid effect	yes (high PTH values)	if PTH values normal/low
from increased serum calcium	yes (high PTH values)	if PTH normal/low
Reverse osteitis fibrosa	yes (if present)	
Reduce bone formation rate	yes, if elevated	cause 'aplastic' bone
Modify immune function	? improvement	uncertain

oral administration, but certain actions have also been studied following the parenteral administration of calcitriol.

Effect on the Intestine

Calcium Absorption. The action of oral calcitriol to increase intestinal calcium absorption has been studied extensively in patients with renal failure; this effect occurs with doses as small as $0.12\ \mu g/day$ [8, 9], and its magnitude increases with doses up to $2.5–5.0\mu g/day$. The relative increment in calcium absorption can be substantial, with increments of fractional absorption from 20% to as high as 45–50% [8, 9]. With this increased absorption, net calcium balance becomes significantly positive in uremic patients: thus, fecal calcium excretion is substantially lowered with either a trivial increase or no change in urinary calcium [10]. Calcium absorption has not been studied after parenteral administration of calcitriol in man.

Phosphate Absorption. This is augmented by oral calcitriol in patients with renal failure [10, 11], but the dose-response relationship has not been established. It should be noted that a substantial fraction of dietary phosphate is absorbed by patients with renal failure even without calcitriol treatment, and the net absorption of phosphate in patients with renal failure is only slightly lower than that of normal subjects [12]. Thus, the increment of phosphate absorption produced by calcitriol is considerably less than that for calcium. The effect of calcitriol on phosphate absorption in normal man has not been studied extensively, but large doses of calcitriol do not augment urinary phosphate excretion in normals [13, 14], an observation which indicates that there is little or no increase in phosphate absorption. This contrasts to the action of calcitriol to enhance calcium absorption in normals and suggests that the absorptive mechanism for phosphate is saturated by low, 'physiologic' doses of the sterol.

Magnesium Absorption. This is only minimally affected by calcitriol treatment, contrasting to its effect on calcium and phosphate absorption. An effect of calcitriol to augment magnesium absorption was shown with use of an intestinal perfusion technique [15], but metabolic balance studies showed no effect on net magnesium absorption despite substantial increases in the net absorption of both calcium and phosphate [16]. This apparent discrepancy may occur because calcitriol's effect on magnesium absorption is small and 'saturated' by low doses. Also, the vitamin D-dependent component of magnesium absorption may comprise a very small fraction of its total net absorption. In vitamin D-deficient animals, the action of calcitriol to enhance magnesium absorption is maximal with a low dose of the sterol in contrast to a wide dose-response relationship for the absorption of calcium and phosphate [17].

Mucosal Hypertrophy. In addition to these effects on intestinal transport, calcitriol treatment produces thickening of the intestinal mucosa in patients with renal failure [18], an effect that is well known in vitamin-D-deficient animals [19]. Certain putative actions of calcitriol on intestinal absorption, such as the effect on aluminum and zinc, could arise because of mucosal hyperplasia rather than from an action on a specific transport mechanism. Based on changes in serum aluminum, it was suggested that 1-α-(hydroxy)-vitamin D might augment aluminum absorption in dialysis patients [20]. However, calcitriol had no effect on aluminum absorption in uremic animals, in which true absorption could be evaluated more accurately; this contrasted to a small positive effect in vitamin D-deficient animals with normal renal function [21]. The action of calcitriol on aluminum absorption in dialysis patients has not been assessed. Since the requirement for phosphate-binding agents may be increased by calcitriol therapy due to augmented phosphate absorption, serum aluminum may rise due to the greater need for aluminum-containing gels in patients receiving vitamin D sterols.

Effect on Serum Calcium

Calcitriol has a major effect to raise the serum calcium level in patients with end-stage renal disease [2–4]; this action can correct the hypocalcemia of renal failure – an important indication for calcitriol therapy in such patients. The mechanism for the increase in serum calcium is multifactorial, but the augmented calcium absorption, discussed above, is a major factor. There is a greater tendency for serum calcium to increase above normal in patients with renal failure because urinary calcium cannot be augmented significantly, in contrast to the effects of calcitriol in normal subjects [13, 14].

Because there is substantial variability in the size of the 'pool' into which the absorbed calcium is distributed, the rise of serum calcium with calcitriol treatment is not dose-related. Despite the fact that calcium absorption is augmented and the net balance for calcium is positive [22], serum calcium may not rise for weeks to even months after calcitriol therapy is initiated in dialysis patients. When this occurs, one can infer that there is substantial deposition of calcium into bone, particularly during the early correction of severe osteitis fibrosa. In patients with low bone turnover and impaired entry of calcium into bone, such as aluminum-related bone disease [23], serum calcium can increase quickly after low doses of calcitriol [24]. A prompt rise in serum calcium level can also occur soon after calcitriol therapy is initiated in patients with marked secondary hyperparathyroidism and high bone turnover [24]; hence, a change in serum calcium during treatment does not predict the type of bone disease that exists. With severe osteitis fibrosa, a rise in serum calcium can occur when calcium absorption is increased in combination with markedly elevated bone resorption. Any effect of calcitriol to lower PTH levels or to reduce bone turnover may be too slow and/or too small to prevent the rise in serum calcium. The adminstration of either oral or parenteral calcitriol can raise the serum calcium level [2, 3, 7], but there are no comparisons utilizing similar doses or intervals between doses. It is the impression that the same dose of intravenous compared to oral calcitriol is less likely to raise serum calcium, but no objective data are available to support this conclusion. Of interest, this effect to raise serum calcium often limits the dose of calcitriol that can be given to many patients with renal failure; therefore, it is often impossible to administer a dose that is equivalent to the normal calcitriol generation rate of $1.0-1.5\ \mu g/day$.

Effect on Other Target Tissues

Bone. The specific effects of calcitriol on bone are not well delineated in patients with end-stage renal disease; this is largely due to a lack of precision in the techniques available for in vivo study. In large doses, calcitriol augments the release of calcium from bone in experimental animals [25], and this effect can occur in humans with vitamin D intoxication [26]; however, this is not well documented in patients with end-stage renal disease.

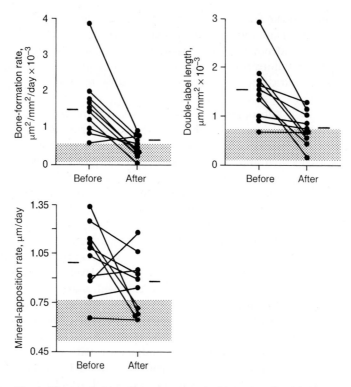

Fig. 1. Changes in bone formation rate, double-tetracycline label length, and mineral apposition rate in 10 patients with severe osteitis fibrosa; values are obtained on bone biopsies before and after therapy with intravenous calcitriol. The mean values are indicated by the horizontal bars; the normal range is indicated by the shaded area. Reprinted from Andress et al. [27], with permission of the publishers.

An important action of calcitriol in patients with osteitis fibrosa has been to reverse many of the features of secondary hyperparathyroidism, with a decrease in the elevated bone formation rate and a reduction of the mineral apposition toward normal. With the administration of both oral and intravenous calcitriol, serum iPTH levels fell, and it is tempting to attribute the changes of histomorphometry of bone to lower PTH levels. Andress et al. [27] reported a reduction in bone formation rate, a decrease in mineralization lag time, reduced peritrabecular fibrosis, and reduced osteoblastic osteoid following intravenous calcitriol (fig. 1); these changes occurred in association with a fall in serum iPTH levels. In reports utilizing oral calcitriol, significant changes in bone have also been observed. Thus, Sherrard et al. [28] noted a significant reduction in peritrabecular fibrosis during therapy with calcitriol in 16 of 17 patients with osteitis fibrosa; also, there was a substantial reduction of unmineralized osteoid in these patients. Malluche et al. [29] reported a significant reduction in endosteal fibrosis in 12 patients treated with oral calcitriol for 6 months; moreover, the percent of actively mineralizing osteoid seams increased in 10 of 11 patients with abnormal pretreatment values, and there was a reduction in woven osteoid in 9 of 12 patients. With oral calcitriol therapy for 12 months, Baker et al. [30] observed a significant change in bone histomorphometry, including reduced lamellar osteoid volume and seam thickness, reduced woven osteoid and a reduction of both bone-osteoblast interface and the osteoblastic index in patients with moderate renal failure despite a lack of statistically significant change of serum iPTH levels (C-terminal). Norris et al. [31] observed a $38 \pm 6\%$ fall in serum alkaline phosphatase before there was any change in serum iPTH levels during intravenous calcitriol therapy in dialysis patients with very severe secondary hyperparathryoidism. Such observations suggest that calcitriol may have an effect on bone, primarily to inhibit osteoblast function, independent of its action to lower PTH levels. It is not possible to compare reported results with oral versus the intravenous route of administration, although Andress et al. [27] achieved their results in several patients who could tolerate only small daily doses of oral calcitriol therapy because of the development of hypercalcemia. There may be some concern regarding whether bone turnover may be reduced to subnormal, creating a state of low bone turnover or 'aplastic' bone disease.

Parathyroid Gland Suppression. When treatment with calcitriol was associated with suppressed serum iPTH levels, this had been attributed to a rise in serum calcium. However, in vivo studies carried out in vitamin-D-deficient puppies [32] and observations in patients with acute renal failure who underwent peritoneal dialysis using a low calcium concentration in dialysate to prevent a rise in serum calcium [33] provided evidence that calcitriol therapy could suppress serum PTH levels directly or potentiate the effect of a higher serum calcium level. Subsequently, both in vivo and in vitro studies have shown that calcitriol acts to inhibit generation of the mRNA for synthesis of pre-, pro-PTH [34, 35]. The results of Slatopolsky et al. [7], described in detail below, in patients with end-stage renal failure provide strong support for the view that intravenous calcitriol directly reduces serum iPTH levels. One difficulty with the interpretation of in vivo studies is separating the effect of calcitriol to raise serum calcium, which in itself can lower PTH secretion, from an independent effect of the sterol to inhibit PTH synthesis. To a clinician, who is interested in the overall effect, the separation is unimportant as long as the degree of hypercalcemia required is not hazardous. To date, no comprehensive studies are available to compare the action of oral compared to intravenous calcitriol. The comparisons in table 2 are limited to results that utilize comparable assays for serum iPTH; this is done because of considerable variability in the reduction of serum iPTH levels depending on the PTH immunoassay employed [27, 36].

Slatopolsky et al. [7] provided evidence that parenteral calcitriol had a substantial effect to lower serum iPTH levels, as measured with a mid-region assay. They gave calcitriol intravenously three times weekly to 20 dialysis patients with hypocalcemia and mild to moderate secondary hyper-parathyroidisim; the dose was increased over 8 weeks from 0.5 to 4.0 μg per treatment and was temporarily discontinued if the serum calcium exceeded 11.5 mg/dl. They observed a 70.1 ± 3.2% decrease in serum iPTH (mid-region) in the total group of patients, but the magnitude of reduction was greatest in those with mildly elevated iPTH levels. Moreover, there was a 20% decrease in iPTH before serum calcium changed. After the therapy with intravenous calcitriol was discontinued, 5 of the 20 patients were given calcium carbonate in progressively larger doses to raise their serum calcium levels to the same value observed during intravenous calcitriol; with similar increments of serum calcium levels, serum iPTH was reduced 73.5 ± 5.1% (SE) from control values by calcitriol, compared to 26 ± 6.7% by calcium carbonate. There was no comparison with oral calcitriol; but data from 3 other dialysis patients with serum iPTH levels that were substantially higher exhibited no change in serum Ca or iPTH during therapy with oral calcitriol, 0.5 μg daily, for 6 months.

In preliminary studies, Norris et al. [31] gave intravenous calcitriol, 1.0–5.0 μg, with each dialysis, to 15 dialysis patients with marked and symptomatic secondary hyperparathyroidism for 13 ± 2 months. Unlike the patients studied by Slatopolsky et al. [7], serum mid-region iPTH levels were markedly elevated (1,250 ± 275 μlEq/ml; normal < 9 μlEq/ml), serum alkaline phosphatase levels were increased to 312 ± 71 IU/l and the patients had normal serum calcium levels. After 13 months of treatment, serum iPTH had decreased by 42 ± 4%, and alkaline phosphatase was reduced by 62 ± 14% while serum calcium rose from 10.0 ± 0.1 to 11.2 ± 0.1 mg/dl. There was a 33 ± 6% decrease in serum iPTH during the early weeks of therapy before serum calcium changed. These observations were qualita-tively similar to those of Slatopolsky et al, but the PTH suppression was quantitatively less and much more time was required before serum iPTH levels fell; the greater severity of secondary hyperparathyroidism probably accounts for the differences, but the mechanism is uncertain.

In studies with oral calcitriol, serum iPTH levels often decreased during the period of therapy, but this effect has generally correlated with the rise in serum calcium. In a double-blind controlled study of oral calcitriol compared to placebo [37], there was a 43% decrease in mid-region serum iPTH levels in dialysis patients receiving oral calcitriol (p = 0.05), compared to a 30% rise in a group receiving placebo (p > 0.05). In another controlled study of 98 dialysis patients without clinical features of bone disease and who were treated with either placebo or calcitriol for 30 months, serum iPTH levels generally fell in the calcitriol-treated group and rose in those receiving placebo [38, 39]. There were marked differences among individual patients: among the calcitriol-treated patients, serum iPTH (mid-region) was significantly suppressed in 60% of the group, iPTH was unchanged in 30%, and it rose in 10%; this contrasts to suppression of iPTH in 27% of the placebo group, no change in iPTH in 34%, and a rise in 39% of the placebo-treated patients (p < 0.01, by χ^2) [39]. These effects occurred as serum calcium rose from 9.4 ± 0.6 to 10.2 ± 0.5 (SD) with calcitriol treatment, and from 9.5 ± 0.6 to 9.7 ± 0.6 mg/dl in the placebo group. In children with renal failure managed with CAPD, 'high' dose oral calcitriol (0.52 μg/day) was given; serum mid-region iPTH fell by 27% as serum calcium rose, and there was improvement of skeletal radiographs and normalization of serum alkaline phosphatase [40]. In a long-term study of adult patients with symptomatic renal osteodystrophy, oral calcitriol was given in an initial dose of 1.0 μg/day and later reduced to 0.25–0.125 μg/day when hypercalcemia developed [41]; treatment was continued for 30–44 months. There were a reduction of serum alkaline phosphatase levels, improvement of bone pain and muscular weakness, but no change in serum iPTH levels, as measured with a mid-region PTH assay.

In comparison of oral versus intravenous calcitriol, Norris et al. [31] carried out a double cross-over study in 3 patients with severe osteitis fibrosa; the patients received thrice weekly intravenous therapy for 3 months, daily oral therapy for 6 months, followed by 3 more months with intravenous calcitriol. Serum calcium and mid-region iPTH levels were measured 1–3 times weekly. Serum iPTH levels were expressed as a percent of the pretreatment value. For all serum calcium levels below 11.0 mg/dl (10.6 ± 0.1 mg/dl during intravenous treatment and 10.7 ± 0.1 with oral), serum iPTH levels were suppressed from the pretreatment value by 45.1 ± 2.0 and 25.6 ± 4.0%, respectively, with intravenous and oral therapy. With all serum calcium levels of 11.0 mg/dl and above (11.4 ± 0.1 and 11.5 ± 0.1 with intravenous and oral, respectively), the iPTH levels were reduced by 62.5 ± 2.0 and 48.6 ± 7.0%, respectively, with intravenous and oral therapy; these changes were significantly different, p < 0.01. Such data suggest that intravenous calcitriol is more effective than daily oral treatment in suppressing iPTH levels with similar levels of serum calcium.

The degree of suppression of serum iPTH levels has generally been more marked during calcitriol therapy when an 'intact' or 'amino-terminal' PTH assay was employed. In a study of 32 patients with symptomatic

Table 2. Effect of calcitriol in patients with end-stage renal disease: summary of selected reports

Ref. No.	Patients treated n	Serum Ca pre-Rx mg/dl	Severity of hyper-parathyroidism	Duration of therapy months	Dose[1] µg/day or µg/dialysis	Serum iPTH assay type	% reduction from baseline	Comment
Oral therapy:								
37	64	<12.0	asymptomatic	24	0.50–0.25	mid-region	−43 in 1,25D +30 in placebo	mild erosions improved
35, 36	98	<11.0	asymptomatic normal X-rays	30	0.50–0.125	mid-region	*fell, 60 of 1,25D rose, 39 of placebo	erosions prevented S-1,25D, 7±7 → 16±5 pg/ml
41	8	<10.4	symptomatic	30–44	1.0–0.125	mid-region	no change	symptoms improved alkaline phosphatase fell
40	18	<10.8	asymptomatic	12–18	0.61±0.09	mid-region	−27	S-Ca, 9.9±0.2 → 11.±0.0.2 children on CAPD
42, 43	32	<11.0	symptomatic Bx: OF, mixed	12–18	average 0.62	amino-terminal	−77	S-Ca, 9.4±0.2 → 10.1±0.3 in OF S-Ca, 7.2±0.4 → 8.8±0.3 in mixed osteitis fibrosa improved
44	8	<10.0	nPTH>normal	15	0.60±0.07	intact	−44±7 at 9 months −71±9 at 15 months	SCa^{2+}, 1.10±0.3 → 1.27±0.04 mM S-1,25D, 7±2 → 21±1.4 pg/ml

Intravenous therapy

7	20	low S-Ca	mild-moderate	2	0.5–4.0	mid-region	−70.1±3.2	S-Ca, 8.5±0.3 → 9.4±0.3 mg/dl
31	15	<11.0	symptomatic severe	13	2.0–5.0	mid-region	−42±4	S-Ca, 10.1±.1 → 11.2±0.1 mg/dl
27	5	<11.5	severe	12	1.0–2.5	amino-terminal	−60	S-Ca, 10.2 → 10.7 mg/dl osteitis fibrosa improved
45	10	ND	nPTH 9 × normal	2.5	3.0	amino-terminal	−22	low Ca dialysate SCa^{2+}, 1.22±0.05 → 1.17±0.04 mM
46	8	ND	severe nPTH 14 × normal	2.5	2.0	intact	−52	low Ca dialysate SCa^{2+}, 1.12±0.03 → 1.10±0.02 mM

S-Ca = Serum calcium; 1.25D = calcitriol-treated group; placebo = placebo-treated group; SCa^{2+} = ionized serum Ca; OF = osteitis fibrosa; Bx = bone biopsy; ND = no data; * = percentage of group showing suppression.
Modified from: Coburn JW: The use of oral and parenteral calcitriol in the treatment of renal osteodystrophy. Kidney Int 1990: (suppl): in press.
^{1} Oral: μg/day; intravenous, μg/dialysis.

secondary hyperparathryoidism and bone biopsies showing osteitis fibrosa or mixed disease, the oral dose of calcitriol averaged 0.62 µg/day; serum iPTH, measured with antiserum 211/32, which reacts with the intact PTH molecule, fell by 77% after 6–8 months of therapy [42, 43]. Quarles et al. [44] measured intact PTH with an immunoradiometric assay (IRMA) for PTH in 8 dialysis patients receiving daily oral calcitriol and calcium carbonate. They noted a fall in serum iPTH by $44 \pm 7\%$ at 9 months and by $71 \pm 9\%$ at 15 months as serum ionized calcium rose from 1.10 ± 0.27 to 1.27 ± 0.04 mmol/l. In these trials, the larger doses of oral calcitriol are emphasized by Quarles et al. [44], with plasma calcitriol levels reaching the low normal range.

In studies with the measurement of intact PTH levels, Delmez et al. [45] gave intravenous calcitriol, 3.0 µg per dialysis for 10 weeks, to 10 patients as dialysis was done with dialysate calcium reduced to 2.5 mEq/l. There was a 22% decrease in serum intact PTH as ionized calcium fell insignificantly from 1.22 ± 0.05 to 1.17 ± 0.04 mmol/l. In another study [46], intravenous calcitriol, 2.0 µg per dialysis, was given for 10 weeks to 9 dialysis patients who underwent dialysis with dialysate calcium lowered to 2.5 mEq/l. There was a 52% decrease in intact PTH while serum ionized calcium did not change significantly (1.12 ± 0.03 pretherapy to 1.10 ± 0.02 mmol/l at the end). Thus, these two short-term studies showed a decrease in intact iPTH levels in the face of no change in serum calcium during intravenous calcitriol.

Immune System. Growing evidence indicates that calcitriol can affect immunoregulation [47, 48], and the calcitriol deficiency of renal insufficiency could contribute to the immunosuppression that often exists [48]. Studies of the effect of calcitriol on immune function in patients with ESRD are limited. In a preliminary study, Bargman et al. [49] found that oral calcitriol therapy of dialysis patients for 14 days enhanced the PHA-stimulated interleukin-2 production by lymphocytes and increased the ratio of helper: suppressor T lymphocytes. In contrast, Klein et al. [50] found variable effects of parenteral calcitriol on lymphokine production and on helper/suppressor ratio in patients studied after 2 and 6 weeks of intravenous calcitriol. The reason for the discrepancy is uncertain; further studies of these effects of calcitriol are needed.

Comparison of Oral versus Intravenous Calcitriol:
Are They Different and Why?

The absence of controlled comparisons of intravenous compared to oral calcitriol makes it difficult to answer the questions of how the treatment modalities differ or why the differences exist. With oral therapy with calcitriol, the doses were traditionally given once or twice daily, while intravenous calcitriol is given thrice or twice per week, depending on the schedule of dialysis treatments. The individual doses of calcitriol and the average dose per week have both been larger with intravenous than oral treatment: thus, the calcitriol doses have averaged 2.0–2.5 μg with each dialysis in the intravenous trials, and these doses extrapolate to 0.86 to 1.1 μg/day over the week. With oral therapy, only a few trials have utilized doses this large [3, 29], and hypercalcemia frequently necessitated a substantial dose reduction, particularly when calcitriol therapy extended to 12 months or longer. With few exceptions [27, 31, 51], the trials with intravenous therapy have not extended for more than 1 year [27, 21]. The production rate of calcitriol is estimated to be 1.0–1.5 μg/day in normal humans [52], and this quantity is rarely tolerated in dialysis patients receiving daily oral therapy.

One reason for the variable effects with the two routes of administration is that oral calcitriol may stimulate intestinal actions to a greater degree than the same intravenous dose; this would produce greater calcium absorption and a propensity to hypercalcemia. If significant degradation of orally administered calcitriol occurs in the gut, the amount absorbed and the rise of plasma calcitriol above the baseline value would be less following an oral than intravenous dose. In a study of 'bioavailability' with the two routes, Salusky et al. [53] measured the 'area under the curve' for plasma calcitriol after separate intravenous and oral doses of the sterol in adolescent patients with end-stage renal disease (fig. 2). The area under the curve was 62% greater after an intravenous dose of calcitriol compared to the bolus oral dose. These data indicate a 'single pass' effect on the gut, with intestinal degradation of part of the oral sterol, and this could account for differences between oral and intravenous administration. This is probably not the only mechanism responsible; if so, one would expect that 1-α-(hydroxy)-vitamin D_3, which undergoes hepatic bioconversion to calcitriol before it exerts a biologic effect, would have advantages over oral calcitriol. The comparisons of actions of these two sterols are limited to short-term actions [9], but reports of long-term effects of 1-α(OH)-vitamin D_3 compared to those of calcitriol suggest no substantial benefit of the former [3], and hypercalcemia has limited the daily dose of 1-α(OH)-vitamin D_3 much like calcitriol. Another difference is the strikingly higher plasma calcitriol levels achieved after an intravenous compared to oral calcitriol [7, 53]. If the inhibition of the mRNA for synthesis of pre-, pro-PTH requires high or even 'supranormal' levels of calcitriol for suppression, the 'high' plasma calcitriol levels achieved after intravenous treatment might account for the difference compared to oral treatment.

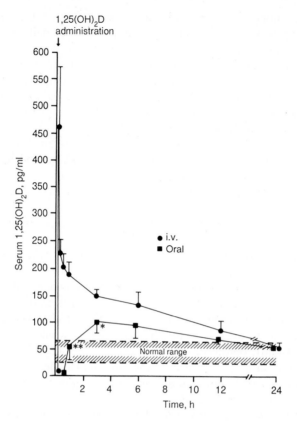

Fig. 2. Changes in serum levels of 1,25(OH)$_2$D in 6 adolescent patients being managed with continuous ambulatory peritoneal dialysis for end-stage renal disease. On two separate occasions, separated by 1 month, they received single oral and intravenous doses of calcitriol at a dose adjusted to 4.0 μg/70 kg body weight. The values are mean ± SE; the values differ: *p < 0.05, **p < 0.01; modified from Salusky et al. [53] and reprinted from Coburn JW: The use of oral and parenteral calcitriol in the treatment of renal osteodystrophy. Kidney Int 1990;(suppl):in press, with permission of the publishers.

Although plasma calcitriol levels of 100–200 pg/ml may seem 'pharmaceutical' rather than 'physiologic', very high levels are generated in vivo when patients with nutritional vitamin D deficiency and normal renal function receive modest doses of vitamin D [54]. Under such circumstances, the plasma calcitriol levels remain substantially elevated for several weeks, perhaps until the accompanying secondary hyperparathyroidism has regressed. If high plasma calcitriol levels are important for PTH suppression, such levels might be achieved after large pulse doses of oral calcitriol. Indeed, preliminary trials provide evidence for the effectiveness of twice or

Table 3. Wholesale cost of calcitriol, oral versus parenteral, 1989

	Preparation	
	intravenous	oral
Dose of 2.0 μg:	1 × 2.0 μg vial	4 × 0.50 μg capsules
Pharmacy A, US$	10.32	3.36
Pharmacy B, US$	10.36	4.64

Wholesale costs quoted by two hospital pharmacies in the Los Angeles, one Federal and the other private. Other costs (dispensing, syringes, administration, etc.) are not included.

thrice weekly oral calcitriol therapy [55, 56]. Moreover, lowering the calcium concentration in dialysate in combination with the oral pulse therapy with 'higher' doses of calcitriol may be another means to permit adequate parathyroid suppression with less risk of hypercalcemia [57].

An important difference between the two forms of therapy is the relatively greater cost of the intravenous form (table 3). Thus, there should be considerable advantage of the intravenous form to justify the additional cost, and careful comparative studies are needed. Presently, we would recommend calcitriol therapy for dialysis patients with evidence of significant secondary hyperparathyroidism. If the patient is hypocalcemic and if 'adequate' doses of oral calcitriol (e.g. doses of at least 0.50 μg/day) can be given without producing hypercalcemia, oral therapy would be recommended first. If the patient's serum calcium level is in the upper normal range, if oral calcitriol produces hypercalcemia, or if there are symptoms of bone disease and parathyroid surgery is being considered, a trial with intravenous calcitriol is warranted providing there are no contraindications for therapy (e.g. marked hyperphosphatemia (serum $p > 7.0$ mg/dl) or a markedly elevated Ca × P product (values above 70–75), for therapy). Comparative studies are required to establish the specific guidelines for such therapy.

References

1   Fraser DR, Kodicek E: Unique biosynthesis by kidney of a biologically active vitamin D metabolite. Nature 1970;228:764–766.
2   Brickman AS, Coburn JW, Norman AW: Effect of 1,25-dihydroxycholecalciferol, the active metabolite of vitamin D, in uremic man. N Engl J Med 1972;287:891–895.
3   Peacock M: The clinical uses of 1-α-hydroxyvitamin D₃. Clin Endocrinol 1977;7(suppl):1S–246S.

4 Coburn JW, Massry SG (ed): Uses and actions of 1,25-dihydroxyvitamin D₃ in uremia. Contr Nephrol. Basel, Karger, 1980, vol 18, pp 1–217.

5 Portale AA, Booth BE, Halloran BP, Morris RC Jr: Effect of dietary phosphorus on circulating concentrations of 1,25-dihydroxyvitamin D and immunoreactive parathyroid hormone in children with moderate renal insufficiency. J Clin Invest 1984;73:1580–1589.

6 Portale AA, Booth BE, Tsai HC, Morris RC Jr: Reduced plasma concentration of 1,25-dihydroxyvitamin D in children with moderate renal insufficiency. Kidney Int 1982;21:627–633.

7 Slatopolsky E, Weerts C, Thielan J, et al: Marked suppression of secondary hyperparathyroidism by intravenous administration of 1,25-dihydroxycholecalciferol in uremic patients. J Clin Invest 1984;74:2136–43.

8 Brickman AS, Coburn JW, Massry SG, et al: 1,25-dihydroxy-vitamin D₃ in normal man and patients with renal failure. Ann Intern Med 1974;80:161–163.

9 Brickman AS, Coburn JW, Friedman GR, et al: Comparison of effects of 1α-hydroxyvitamin D₃ and 1,25-dihydroxyvitamin D₃ in man. J Clin Invest 1976;57:1540–1547.

10 Brickman AS, Hartenbower DL, Norman AW, et al: Actions of 1α-hydroxy- and 1,25-dihydroxyvitamin D₃ on mineral metabolism in man. I. Effects on net absorption of phosphorus. Am J Clin Nutr 1977;30:1064–1070.

11 Ramirez JA, Emmett M, White MG, et al: The absorption of dietary phosphorus and calcium in hemodialysis patients. Kidney Int 1986;30:753–759.

12 Coburn JW, Hartenbower DL, Brickman AS, et al: Intestinal absorption of calcium, magnesium and phosphorus in chronic renal insufficiency; in David DS (ed): Calcium Metabolism in Renal Failure and Nephrolithiasis. New York, Wiley, 1977, pp 77–109.

13 Levine BS, Singer FR, Bryce GF, et al: Pharmacokinetics and biochemical effects of calcitriol in normal man. J Lab Clin Med 1985;106:239–246.

14 Smothers RL, Levine BS, Singer FR, et al: The relationship between urinary calcium and calcium intake during treatment with calcitriol. Kidney Int 1986;29:578–583.

15 Schmulen AC, Lerman M, Pak CYC, et al: Effect of 1,25-dihydroxyvitamin D₃ therapy on jejunal absorption of magnesium in patients with chronic renal failure. Am J Physiol 1980;238:349G–355G.

16 Coburn JW, Brickman AS, Hartenbower DL, Norman AW: Effect of 1,25-dihydroxyvitamin D and 1α-hydroxyvitamin D on magnesium metabolism in man; in Cantin M, Seelig M (eds): Magnesium in Health and Disease. Jamaica, NY, Spectrum, 1980, pp 268–273.

17 Levine BS, Walling MW, Coburn JW: Effect of vitamin D sterols and dietary magnesium on calcium and phosphorous homeostasis. Am J Physiol 1981;241:E35–E41.

18 Goldstein DA, Horwitz RE, Petit S, et al: The duodenal mucosa in patients with renal failure: Response to 1,25(OH)₂D₃. Kidney Int 1981;19:324–331.

19 Spielvogel AM, Farley RD, Norman AW: Studies on the mechanism of action of calciferol. Exp Cell Res 1972;74:359–66.

20 Demontis R, Leflon A, Fournier A, et al: 1α(OH) vitamin D₃ increases plasma aluminum in hemodialyzed patients taking Al(OH)₃. Clin Nephol 1986;26:146–149.

21 Ittel TH, Kluge R, Sieberth HG: Enhanced gastrointestinal absorption of aluminum in uremia: Time course and effect of vitamin D. Nephrol Dialysis Transplant 1986;3:617–23.

22 Coburn JW, Brickman AS: Current status of the use of newer analogs of vitamin D in the management of renal osteodystrophy; in Massry SG, Ritz E, Rapado A (eds): Homeostasis of Phosphate and Other Minerals. New York, Plenum Press, 1978, pp 473–486.

23 Alfrey AC: The case against aluminum affecting parathyroid function. Am J Kidney Dis 1984;6:309–312.

24   Coburn JW, AS Brickman, DJ Sherrard, et al: Use of 1,25(OH)$_2$-vitamin D$_3$ to separate 'types' of renal osteodystrophy. Proc Eur Dialysis Transplant Assoc 1977;14:442–450.

25   Reynolds JJ, Holick MF, DeLuca HF: The role of vitamin D metabolites in bone resorption. Calcif Tissue Res 1973;12:295–301.

26   Coburn JW, Barbour G: Vitamin D intoxication and sarcoidosis; in Coe F (ed): Hypercalciuric States: Pathogenesis, Consequences, and Treatment. Orlando, Grunne & Stratton, 1984, pp 379–433.

27   Andress DL, Norris KC, Coburn JW, Slatopolsky E, Sherrard DJ: Intravenous calcitriol in the treatment of refractory osteitis fibrosa of chronic renal failure. N Engl J Med 1980;321:274–279.

28   Sherrard DJ, Coburn JW, Brickman AS, et al: Skeletal response to treatment with 1,25-dihydroxyvitamin D in renal failure. Contr Nephrol. Basel, Karger, 1980, vol 18, pp 92–97.

29   Malluche HH, Goldstein DA, Malluche SG: Effects of 6 months therapy with 1,25(OH)$_2$D$_3$ on bone disease of dialysis patients. Contr Nephrol. Basel, Karger, 1980, vol 18, pp 98–104.

30   Barker LRI, Abrams L, Roe CJ, et al: 1,25(OH)$_2$D$_3$ administration in moderate renal failure: A prospective double-blind trial. Kidney Int 1989;35:661–69.

31   Norris KC, Kraut JA, Andress DL, et al: Intravenous calcitriol: Effects in severe secondary hyperparathyroidism. J Bone Miner Res 1986;1(suppl 1):374.

32   Oldham SB, Smith R, Hartenbower DL, et al: The acute effects of 1,25-dihydroxycholecalciferol on serum immunoreactive parathyroid hormone in the dog. Endocrinology 1979;104:248–254.

33   Madsen S, Olgaard K, Ladefoged J: Suppressive effect of 1,25-dihydroxyvitamin D$_3$ on circulationg parathyroid hormone in renal failure. J Clin Endocrinol Metab 1981;53:823–827.

34   Russell J, Lettieri D, Sherwood LM: Suppression by 1,25(OH)$_2$D$_3$ of transcription of the parathyroid hormone gene. Endocrinology 1986;119:2864–2866.

35   Silver J, Naveh-Many T, Mayer H, et al: Regulation by vitamin D metabolites of parathyroid hormone gene in vivo by the rat. J Clin Invest 1986;78:1296–1301.

36   Bellazi R, Romanini D, de Vincenzi A, Volpini T: Secondary hyperparathyroidism, anemia and treatment with 1,25(OH)$_2$D$_3$. Nephron 1987;45:251–52.

37   Memmos DE, Eastwood JB, Talner LB, et al: Double-blind trial of oral 1,25-dihydroxy vitamin D$_3$ versus placebo in asymptomatic hyperparathyroidism in patients receiving maintenance haemodialysis. Br Med J 1981;282:1919–1924.

38   Coburn JW, DiDomenico NC, Bryce GF, et al: Prospective, double blind trial with calcitriol in the prophlaxis bone disease in asymptomatic dialysis patients; in Norman AW, Schaefer K, Grigoleit HG, et al (eds): Vitamin D: Chemical, Biochemical, and Clinical Endocrinology of Calcium Metabolism. Berlin, DeGruyter, 1982, pp 833–834.

39   Coburn JW, DiDomenico NC, Bryce GF, et al: Use of calcitriol in prophylaxis of bone disease in dialysis patients: A prospective, double blind study. Kidney Int 1983;23:145.

40   Salusky IB, Fine RN, Kangarloo H, et al: 'High-dose' calcitriol for control of renal osteodystrophy in children on CAPD. Kidney Int 1987;32:89–95.

41   Moorthy AV, Harrington AR, Mazess RB, Simpson DP: Long-term therapy of uremic osteodystrophy in adults with calcitriol. Clin Nephrol 1981;16:93–100.

42   Coburn JW, Brickman AS, Sherrard DJ, et al: Clinical efficacy of 1,25-dihydroxyvitamin D$_3$ in renal osteodystrophy; in Norman AW, Schaefer K, Coburn JW, et al (eds): Vitamin D: Biochemical, Chemical and Clinical Aspects Related to Calcium Metabolism. Berlin, de Gruyter, 1977, pp 657–666.

43 Coburn JW, Sherrard DJ, Ott SA, et al: Use of active vitamin D sterols in end-stage renal failure; in Caniggia A (ed): Proc 4th Int Congr Calciotropic Hormones. Genova, Medical Systems SpA, 1984, pp 27–33.

44 Quarles LD, Davidai GA, Schwab SJ, et al: Oral calcitriol and calcium: Efficient therapy for uremic hyperparathyroidism. Kidney Int 1988;34:840–44.

45 Delmez JA, Tindira C, Grooms P, et al: Parathyroid suppression by intravenous 1,25-dihydroxyvitamin D: A role for increased sensitivity to calcium. J Clin Invest 1989;83:1349–55.

46 Dunlay R, Rodriguez M, Felsenfeld AJ, Llach F: Direct inhibitory effect of calcitriol on parathyroid function (sigmoidal curve) in dialysis patients. Kidney Int 1989;36:1093–98.

47 Tsoukas CK, Provvedini DM, Manolagas SC: 1,25-Dihydroxyvitamin D_3: A novel immunoregulatory hormone. Science 1984;224:1438–40.

48 Reichel H, Koeffler HP, Norman AW: The role of the vitamin D endocrine system in health and disease. N Engl J Med 1989;320:980–991.

49 Bargman JM, Kuzniak S, Klein MH: Changes in immune function induced by 1,25-dihydroxyvitamin D_3. Kidney Int 1987;31:342.

50 Klein GC, Norris KC, Forman SJ, et al: Effect of parenteral $1,25(OH)_2$-vitamin D on lymphocyte lymphocyte proliferation and T lymphocyte helper-suppressor ratio in dialysis patients. Int J Lab Clin Stud Horm Metabol 1987;1:15–21.

51 Trachtman H, Gauthier B: Parenteral calcitriol for treatment of severe renal osteodystrophy in children with chronic renal insufficiency . J Pediatr 1987;110:966–70.

52 Portale AA, Haloran BP, Murphy MM, Morris RC, Jr: Oral intake of phosphorus can determine the serum concentration of 1,25-dihydroxyvitamin D by determining its production rate in humans. J Clin Invest 1986;77:7–12.

53 Salusky IB, Goodman WG, Norris KC, et al: Bioavailability of calcitriol after oral, intravenous, and intraperitoneal doses in dialysis patients; in Norman AW, Schaefer K, Grigoleit HG, v Herrath D (eds): Vitamin D. Molecular, Cellular and Clinical Endocrinology. Berlin, Walter de Gruyter, 1988, pp 783–784.

54 Papapoulos SE, Fraher LJ, Clemens TL, et al : Metabolites of vitamin D in human vitamin-D deficiency: Effect of vitamin D_3 or 1,25 dihydroxycholecalciferol. Lancet 1980;ii:612–615.

55 Martin KJ, Ballal S, Domoto D, et al: Pulse oral calcitriol for the treatment of hyperparathyroidism in patients on CAPD; in: Abstract Book, Symposium on Renal Bone Disease, Parathyroid Hormone and Vitamin D, Singapore, 1990, p 82.

56 Kitaoka M, Fukagawa M, Takano K, et al: Regression of parathyroid gland hyperplasia by $1,25(OH)_2D_3$ pulse therapy in chronic dialysis patients; in: Abstract Book, Symposium on Renal Bone Disease, Parathyroid Hormone and Vitamin D, Singapore, 1990, p 14.

57 Van Der Merwe WM, Rodger RSC, Grant AC et al: Low calcium dialysate and high dose oral calcitriol treatment of secondary hyperparathyroidism in haemodialysis patients; in: Abstract Book, Symposium on Renal Bone Disease, Parathyroid Hormone and Vitamin D, Singapore, 1990, p 81.

Jack W. Coburn, MD, Nephrology Section (W111L), West Los Angeles
Veterans Affairs Medical Center, Wilshire and Sawtelle Boulevards,
Los Angeles, CA 90073 (USA)

Morii H (ed): Calcium-Regulating Hormones. I. Role in Disease and Aging.
Contrib Nephrol. Basel, Karger, 1991, vol 90, pp 183–188

Treatment with 1,25(OH)$_2$D$_3$ in Predialysis Chronic Renal Failure[1]

G. Coen, S. Mazzaferro, P. Ballanti, S. Costantini, E. Bonucci,
F. Bondatti, M. Manni, M. Pasquali, D. Sardella, F. Taggi

Chair of Nephrology, 'La Sapienza' University, and Istituto Superiore di Sanità,
Rome, Italy

In chronic renal failure there is an early occurrence of secondary hyperparathyroidism which has been attributed to phosphate retention according to the well-known 'trade off' hypothesis [1]. Recently, this view has been subjected to criticism and on the basis of reports of relatively early reduction in 1,25(OH)$_2$D$_3$ serum levels in CRF [2], it has been postulated that decreased production of this hormone is the major event leading to parathyroid hypersecretion and hyperplasia [3, 4]. Therefore, if an early deficit of the sterol is responsible for secondary hyperparathyroidism, prevention and treatment of this condition should be based on the administration since early stages of 1,25(OH)$_2$D$_3$ or other analogues. Since the 1970s, these sterols were employed for the treatment of hyperparathyroidism. The results on bone lesions were satisfactory. However, some authors have warned against an indiscriminate use of the sterol due to possible adverse side effects mainly on renal function [5]. Most of the reports confirming adverse effects of calcitriol on renal function were based on treatment trials employing relatively elevated doses of calcitriol, with episodes of hypercalcemia. Therefore, we decided to start a trial with a lower dose of calcitriol unable to provoke hypercalcemia. The study comprised bone histology [6] and the effects of treatment on the rate of fall of renal function. In addition, considering that there are controversial opinions on the possible influence of vitamin D on aluminum absorption and bone deposition [7, 8], the study has been extended to the effect of treatment on bone aluminum content [9].

A first study was performed on 38 patients with slowly evolving CRF from the out-patient renal clinic of the Nephrology Unit. The patients, 23 males and 15 females, had mean age of 46 ± 16.7 years. They were divided

[1] This study was supported by funds of Ministero della Pubblica Istruzione and of the 'La Sapienza' University of Rome.

Table 1. Biochemical data from 15 patients before and at the end of treatment with $1,25(OH)_2D_3$ (mean ± SD)

	Before	After	p
Crs, mg/dl	4.93 ± 1.70	7.37 ± 3.40	<0.0025
CCr, ml/min	18.45 ± 9.10	12.74 ± 6.16	<0.0025
Cas, mg/dl	9.10 ± 0.37	9.43 ± 0.78	<0.05
Ps, mg/dl	3.95 ± 0.96	4.60 ± 1.10	<0.025
AP, mU/ml	207.23 ± 162.86	126.40 ± 109.46	<0.0005
iPTH, ng/ml	2.04 ± 1.49	1.90 ± 0.92	n.s.
BGP, ng/ml	22.11 ± 9.40	18.49 ± 9.58	n.s.
Cau, mg/day	56.59 ± 30.25	63.75 ± 33.24	n.s.

Table 2. Main histomorphometric parameters of 38 patients with CRF (mean ± SD)

	$1,25(OH)_2D_3$ (n = 15)	Controls (n = 23)	p	Normal values (n = 57)
BV/TV, %	24.35 ± 4.04	23.40 ± 5.55	n.s.	20.33 ± 4.67
OV/BV, %	4.11 ± 4.03	6.38 ± 6.47	n.s.	1.44 ± 1.24
OS/BS, %	26.09 ± 23.75	34.48 ± 24.37	n.s.	9.47 ± 7.30
ObS/BS, %	1.36 ± 2.00	3.36 ± 3.39	<0.05	0.27 ± 0.64
ES/BS, %	2.30 ± 2.37	5.63 ± 4.76	<0.02	1.77 ± 1.50
OcS/BS, %	0.24 ± 0.27	1.11 ± 1.22	<0.01	0.19 ± 0.30

into two groups of comparable rate of decline of renal function: one of 15 patients, mean age 51.2 ± 16.9 years, serum creatinine 4.93 ± 1.7 mg/dl who were treated with $1,25(OH)_2D_3$, 0.25 µg daily for an average period of 16.2 ± 11.3 months; the other of 23 patients mean age 42.3 ± 17.7 years who did not receive vitamin D metabolites and served as controls. At the end of the observation period all patients were subjected to transiliac bone biopsy for histomorphometry. In the treated patients regular controls of serum calcium levels showed values within the normal range throughout the observation period. The results of humoral parameters from the 15 patients before and at the end of treatment are reported in table 1, while histomorphometric parameters of the two groups are reported in table 2. A not significant decrease of iPTH and BGP were observed following treatment. As for the histomorphometric parameters, a significant difference in ObS/BS, ES/BS and OcS/BS was found between the two groups. In conclusion, the treatment was able to improve the bone lesion while stabilizing parathyroid hormone secretion, with no sign of acceleration of the rate of fall of renal function.

Table 3. Clinical and humoral parameters before and at the end of treatment with 1-alpha-OHD$_3$ (n = 10); mean ± SD

	Before	After	p
Crs, mg/dl	4.88 ± 1.49	7.39 ± 3.49	<0.01
Cas, mg/dl	9.04 ± 0.24	9.19 ± 0.57	n.s.
Ps, mg/dl	4.04 ± 0.73	5.26 ± 1.85	<0.02
PTH, ng/ml	1.46 ± 1.04	1.45 ± 1.33	n.s.
BGP, ng/ml	20.34 ± 15.45	18.91 ± 13.82	n.s.
AP, mU/ml	96.05 ± 39.34	68.43 ± 20.29	<0.01

Table 4. Main histomorphometric parameters before and following treatment with 1-alpha-OHD$_3$ (n = 10); mean ± SD

	Before	After	p
OV/TV, %	8.21 ± 7.61	2.21 ± 2.92	<0.002
OS/BS, %	37.71 ± 28.69	14.78 ± 17.22	<0.03
ObS/BS, %	2.17 ± 2.24	0.94 ± 2.24	<0.07
ES/BS, %	5.25 ± 3.79	1.58 ± 2.64	<0.003
OcS/BS, %	0.73 ± 0.73	0.19 ± 0.32	<0.02
N.Oc/TA, n/mm^2	0.66 ± 0.72	0.14 ± 0.26	<0.01

However, the study did not allow to compare the histologic parameters following treatment, with those obtained in basal conditions. Therefore, a separate investigation was run on 10 patients (5 M, 5 F) with slowly evolving chronic renal failure, mean age 42.2 ± 16.5 years, serum creatinine 4.8 ± 1.5 mg/dl. Treatment was carried out with 1-alpha-OHD$_3$, 0.5 μg daily for an average period of 12 months. Transiliac bone biopsies were taken prior to and following treatment. The results of the administration of the analogue on humoral parameters are reported in table 3 while the results on bone histomorphometric parameters are shown in table 4. The humoral parameters were substantially comparable to those obtained in the first study. Treatment induced a significant fall in several histomorphometric parameters, like OV/BV, OS/BS, ES/BS, and a significant increase in BFR-BMU and MF. This study confirmed the favorable results of treatment with small doses of 1.25(OH)$_2$D$_3$ or its analogue 1-alpha-OHD$_3$. However, treatment was not able to reverse to normal the levels of iPTH.

A relevant problem in the treatment with vitamin D metabolites of patients with chronic renal failure may originate from a possible enhancing effect of calcitriol on the tissue aluminum accumulation. Therefore, we carried out a study on two comparable groups of patients treated or

Table 5. Biochemical parameters, serum and bone Al in treated and untreated patients (mean ±SD)

	1,25(OH)$_2$D$_3$	Untreated	p
Crs, mg/dl	6.88 ± 3.33	6.13 ± 2.51	n.s.
Cas, mg/dl	9.40 ± 0.67	8.87 ± 0.72	<0.025
Ps, mg/dl	4.48 ± 1.18	4.35 ± 1.36	n.s.
AP, mU/ml	85.18 ± 38.33	111.12 ± 55.11	n.s.
iPTH, ng/ml	1.47 ± 0.85	3.08 ± 2.43	<0.01
BGP, ng/ml	15.13 ± 8.65	25.72 ± 12.19	<0.005
Als, μg/l	5.54 ± 2.62	5.86 ± 2.85	n.s.
Alb, mg/kg/DW	5.63 ± 4.01	9.59 ± 7.10	<0.05

untreated with calcitriol in order to define the role of the sterol at usual therapeutic doses in the process of aluminum accumulation in bone.

The study has been carried out on a total of 32 patients with slowly evolving chronic renal failure on conservative treatment. They followed a diet moderately restricted in proteins and phosphate with a caloric content of 30–35 cal/kg. None of the patients had ever received aluminum-containing antacids, since their serum phosphate was constantly at satisfactory levels. The mean age of the patients was 56 ± 12.7 years, and the mean serum creatinine was 6.51 ± 2.92 mg/dl. Sixteen of the patients had been treated for the last 13.5 ± 6.7 months with 1,25(OH)$_2$D$_3$, 0.25 μg daily, while the other 16 did not receive the sterol. Following double tetracyclin labeling all patients were subjected to transiliac bone biopsy for histomorphometry and bone aluminum determination. Humoral parameters of the two groups together with serum and bone aluminum content are reported in table 5. It is evident that treatment with 1,25(OH)$_2$D$_3$ did not induce accumulation of bone aluminum which on the contrary was found to be significantly lower than in control subjects. The average value of bone aluminum was lower than in untreated patients in spite of the finding of an increased mineral apposition rate (0.641 ± 0.215 vs. 0.485 ± 0.165 μm/day. The results are in favor of a protective effect of calcitriol on bone from aluminum accumulation.

In the study of the effects of calcitriol in predialysis chronic renal failure our interest was also focused on the possible adverse results on renal function. This problem has been studied on a total of 60 patients. They were randomly assigned to two equal groups, one (A) treated with calcitriol and the other (B) not receiving the sterol. Patients of group A had an average age of 51.9 ± 15.9 years, 17 males and 13 females, 14 with CGN, 12 with TIN, 3 with PKD and 1 with undefined diagnosis. Creatinine

Table 6. Effect of 1,25(OH)$_2$D$_3$ on the slopes of 1/Cr in CRF

Patients	Improved	Stable	Worsened
Treated (a)	14	14	2
(b)	8	21	1
Controls	0	28	2

(a) Comparison between linear regression identified from the start of treatment.
(b) Regression analysis, linear versus parabolic hypothesis.

clearance was 23.4 ± 9 ml/min. Before starting treatment the patients were followed for a period of 15.6 ± 6.3 months. Treatment with calcitriol, 0.25 µg daily, was given on average for a period of 13.6 ± 6.5 months. The control group, B, had average age of 46.2 ± 16.7 years. Sixteen had CGN, 11 TIN and 3 PKD. Creatinine clearance was 27.4 ± 13.3 ml/min. They were followed for an average period of 22.8 ± 10.3 months. Serum creatinine, BUN, calcium, phosphate, iPTH, and urine creatinine, calcium and phosphate were measured periodically during the observation period. Reciprocals of serum creatinine values were calculated. The statistical analysis utilized different techniques, like comparison of linear and parabolic regression analysis with identification of the 'breakpoint' [10]. In addition, data were also analyzed by comparing the regressions obtained after dichotomizing the experimental data at the date of start of treatment.

The results of treatment on serum calcium, phosphate and iPTH were not different from those obtained in the preceding experiments. No episodes of hypercalcemia were recorded during treatment with the sterol. As for the values of urine creatinine before and during treatment, they were $1,261 \pm 310$ and $1,175 \pm 255$ mg/24 h, respectively. Mean arterial pressure and serum BUN/Cr did not change during treatment from basal values. Also in the control group there was a decrease of the same size of urine creatinine (from $1,327 \pm 401$ to $1,233 \pm 347$ mg/24 h), while MAP and urine BUN/Cr did not change during the observation period. In table 6, the behavior of the slopes of serum creatinine reciprocals vs. time are reported. Evaluation of data with comparison of pre- and posttreatment slopes, showed an improvement in the rate of fall of renal function in 14/30 cases. The more restrictive analysis with identification of the breakpoint has shown an improvement of the 1/Cr slope in 8 cases, while no case was identified with this characteristic in the control group. Therefore, these data point to a favorable effect of administration of calcitriol in a percentage of cases ranging from 26 to 46%. These data do not seem to be dependent on confounding factors, since MAP and average dietary protein intake did not change, and the decrease in urine creatinine over time did not differ from control subjects.

In conclusion, $1,25(OH)_2D_3$ at doses not inducing hypercalcemia improves and/or prevents the bone lesion of renal osteodystrophy and the evolution of secondary hyperparathyroidism. Treatment does not enhance bone aluminum accumulation at least in conditions of low exposure to the element. In addition, administration of the sterol in nondialytic renal failure may improve the rate of progression of renal insufficiency. The mechanism underlying this phenomenon requires investigation.

References

1   Bricker RE, Slatopolsky E, Reiss E, Avioli LV: Calcium, phosphorus and bone in renal disease and transplantation. Arch Intern Med 1969;123:543–553.

2   Portale AA, Booth EB, Tsai HC, Morris RC: Reduced plasma concentration of 1,25-dihydroxyvitamin D in children with moderate renal insufficiency. Kidney Int 1982;21:627–632.

3   Wilson L, Felsenfeld A, Drezner MK, Llach F: Altered divalent ion metabolism in early renal failure: Role of 1,25(OH)2D₃. Kidney Int 1985;27:565–573.

4   Szabo A, Merke J, Beier E, Mall G, Ritz E: 1,25(OH)2-vitamin D₃ inhibits parathyroid cell proliferation in experimental uremia. Kidney Int 1989;35:1049–1056.

5   Christiansen C, Rodbro P, Christensen MS, Hartnack B, Transbol I: Deterioration of renal function during treatment of chronic renal failure with 1,25dihydroxyvitamin D₃. Lancet 1978;ii:700–703.

6   Coen G, Mazzaferro S, Bonucci E, Ballanti P, Massimetti C, Donato G, Landi A, Smacchi A, Della Rocca C, Cinotti GA, Taggi F: Treatment of secondary hyperparathyroidism of predialysis chronic renal failure with low doses of 1,25(OH)2D₃: Humoral and histomorphometric results. Mineral Electrolyte Metab 1986;12:375–382.

7   Morinière P, Fournier A, Leflon A, Herve M, Seber JL, Gregoire I, Bataille P, Gueris J: Comparison of 1-alphaOHvitamin D₃ and high doses of calcium carbonate for the control of hyperparathyroidism and hyperalluminemia in patients on maintenance dialysis. Nephron 1985;39:309–315.

8   Hirschberg R, von Herrath D, Voss R, Bossaller W, Mauelshagen U, Pauls A, Schaefer K: Organ distribution of aluminum in uremic rats: Influence of parathyroid hormone and 1,25dihydroxyvitamin D₃. Mineral Electrolyte Metab 1985;11:106–110.

9   Coen G, Mazzaferro S, Costantini S, Ballanti P, Carrieri MP, Giordano R, Smacchi A, Sardella D, Bonucci E, Taggi F: Bone aluminum content in predialysis chronic renal failure and its relation with secondary hyperparathyroidism and 1,25(OH)2D₃ treatment. Mineral Electrolyte Metab 1989;15:295–302.

10   Jones RH, Molitoris BA: A statistical method for determining the breakpoint of two lines. Analyt Biochem 1984;141:287–290.

Prof. Giorgio Coen, Istituto 2°, Clinica Medica, 'La Sapienza' University,
Viale del Policlinico, I–00161 Rome (Italy)

Morii H (ed): Calcium-Regulating Hormones. I. Role in Disease and Aging.
Contrib Nephrol. Basel, Karger, 1991, vol 90, pp 189–195

Evolution of Secondary Hyperparathyroidism during Oral Calcitriol Therapy in Pediatric Renal Osteodystrophy

William G. Goodman, Isidro B. Salusky

Departments of Medicine and Pediatrics, UCLA School of Medicine, Los Angeles, and the Medical and Research Services, Sepulveda Veterans Administration Medical Center, Sepulveda, Calif., USA

Treatment with daily doses of oral calcitriol is effective for the clinical management of secondary hyperparathyroidism in many patients undergoing regular dialysis [1–3]. However, the percentage of patients who demonstrate a favorable response to oral calcitriol as documented by improvements in the skeletal changes of hyperparathyroidism has not been established. Most available data have been obtained in adult patients receiving long-term hemodialysis whereas little information has been reported in patients undergoing chronic peritoneal dialysis, particularly children [4, 5]. Because continuous ambulatory peritoneal dialysis (CAPD) and continuous cycling peritoneal dialysis (CCPD) have been extensively used for the management of pediatric patients with chronic renal failure, additional information is needed regarding the efficacy of daily oral calcitriol therapy in this group of patients.

During the past 6 years, 59 children receiving long-term peritoneal dialysis at the UCLA Medical Center have undergone systematic evaluation, including iliac crest bone biopsy, to characterize the features of renal osteodystrophy in this patient population. The current report summarizes the histologic and biochemical response to 12 months of daily oral calcitriol therapy in patients with established secondary hyperparathyroidism undergoing CAPD/CCPD.

Methods

All patients were clinically stable, and they were treated with either CAPD or CCPD as previously described [6]. The mean age of the patients was 13 ± 6 years, range 4–18 years, and they had been treated with CAPD/CCPD for an average of 17 ± 12 months prior to study. All study protocols were approved by the UCLA Human Subjects Protection Committee, and informed consent was obtained from each patient and/or his or her parents.

All patients were receiving oral calcitriol at the time of initial evaluation. During the study, oral calcitriol was continued, but dosages were adjusted based upon frequent and regular measurements of serum calcium and phosphorus levels [7]. The daily dose of calcitriol was increased in increments of 0.25 μg at monthly intervals to maintain total serum calcium levels between 10.5 and 11.5 mg/dl if serum phosphorus values remained at or below 6.0 mg/dl and serum calcium levels did not exceed 10.5 mg/dl [7].

Either calcium carbonate or aluminum hydroxide was used as the primary phosphate-binding agent. Doses of calcium carbonate ranged from 2.5 to 12 g/day, and these were adjusted according to changes in serum phosphorus levels. Similarly, the daily dose of aluminum hydroxide was adjusted on the basis of serum phosphorus concentrations, but the maximum daily dose of aluminum hydroxide was limited to 30 mg/kg/day as recommended by Sedman et al. [8] for pediatric patients with renal failure. Each phosphate-binding agent was given orally in divided doses with meals.

Samples of blood were obtained at the time of initial bone biopsy and at monthly intervals thereafter for determinations of serum calcium, phosphorus and alkaline phosphatase levels [9]. Serum immunoreactive parathyroid hormone (iPTH) was measured every 3 months using a radioimmunoassay that reacts with the mid-carboxyterminal region of the PTH molecule [9]. Deferoxamine (DFO) infusion tests were done as previously described to assess the degree of aluminum retention in tissues using an intravenous dose of 40 mg/kg body weight [9]; DFO infusion tests were completed before and after 12 months of treatment.

Iliac crest bone biopsies were performed after double tetracycline labeling in each patient prior to and at the completion of study. Quantitative histomorphometry of bone was done as previously reported from this laboratory, and the aurine tricarboxylic acid method was used for the histochemical assessment of aluminum deposition in bone [9]. Reference values for all histologic variables and for tetracycline-based measurements of bone formation in pediatric patients were determined in bone biopsies of the iliac crest obtained from 16 normal children between the ages of 2.5 and 17 years undergoing elective orthopedic or urologic procedures [9]; none of these patients had clinical or biochemical evidence of metabolic bone disease. Pediatric patients with renal osteodystrophy were classified by histomorphometric criteria as osteitis fibrosa, mild lesions of secondary hyperparathyroidism, normal bone histology, aplastic (adynamic) bone, osteomalacia, or mixed lesions of renal osteodystrophy [9].

All results are expressed as the mean ± standard error (SE). Statistical analysis of the data was done using analysis of variance with contrasts and the t test for paired and unpaired samples [10]. The rank-sum test and the sign test were used to assess results that were not normally distributed [10]. Correlation analysis was done by the least-squares method [10].

Results

Of 59 pediatric patients who underwent initial bone biopsy evaluation, 38 had skeletal lesions of secondary hyperparathyroidism (fig. 1); 24 patients had bone biopsy findings of overt osteitis fibrosa whereas 14 demonstrated changes of mild hyperparathyroidism. Of the remaining patients, 10 had normal bone histology and bone formation, 5 had aplastic lesions, 5 had osteomalacia and 1 had a mixed lesion of renal osteodystrophy (fig. 1). Overall, the serum levels of iPTH were higher in patients with overt osteitis

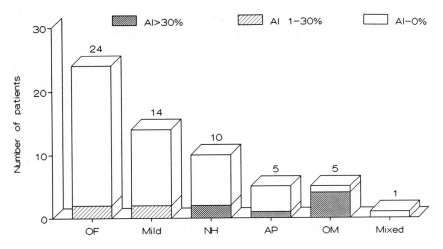

Fig. 1. The distribution of histologic lesions of renal osteodystrophy in 59 pediatric patients undergoing CAPD/CCPD. OF = Osteitis fibrosa; Mild = mild secondary hyperparathyroidism; NH = normal histology; AP = aplastic; OM = osteomalacia; Mixed = mixed lesion. Shaded areas indicate patients with more than 30% bone surface stainable and hatched bars indicate patients with surface stainable aluminum between 1 and 30%.

fibrosa than in those with other histologic lesions of renal osteodystrophy, and serum iPTH correlated with both percent bone resorption surface, $r = 0.48$, $p < 0.05$, and bone formation rate, $r = 0.47$, $p < 0.05$. There was substantial overlap, however, in the serum levels of iPTH among patients with overt osteitis fibrosa and those with other skeletal lesions of renal osteodystrophy.

Seven patients had aluminum-related bone disease as documented by aluminum deposition at more than 30% of trabecular bone surfaces and rates of bone formation that were normal or reduced (fig. 1). Surface stainable aluminum in bone was detected in only 4 patients with osteitis fibrosa or mild lesions, and values were below 30% in each case (fig. 1). Neither basal serum aluminum levels nor the results of DFO infusion tests served to distinguish patients with aluminum-related bone disease from those with other histologic lesions of renal osteodystrophy. Thus, approximately 65% of pediatric patients undergoing CAPD/CCPD have evidence of secondary hyperparathyroidism despite prior treatment with oral calcitriol whereas fewer patients have aluminum-related lesions of bone.

Repeat bone biopsies were obtained after an additional 12 months of intensified therapy with oral calcitriol in 33 patients. Despite this intervention, 18 of 33 patients, or 55%, had skeletal lesions of secondary hyperparathyroidism; 13 patients had osteitis fibrosa and 5 patients had mild lesions. Thus, 1 year of daily oral calcitriol therapy did not substantially alter the prevalence of hyperparathyroid bone disease in pediatric patients undergoing CAPD/CCPD.

Of the 13 patients with osteitis fibrosa on initial bone biopsy, 10 had persistent fibrotic changes on repeat bone biopsy; in 8 of these, the severity of osteitis fibrosa either did not improve or progressed as judged by histomorphometric criteria. Modest improvements in the severity of osteitis fibrosa was observed in 2 patients, each of whom exhibited reductions in the rate of bone formation and in the extent of tissue fibrosis, whereas fibrotic changes resolved completely in 3 patients.

In contrast to these findings, 4 of 6 patients with mild lesions of secondary hyperparathyroidism demonstrated histologic improvement during oral calcitriol therapy; in 3 patients, bone histology and bone formation became normal after 12 months of treatment whereas 1 other patient experienced a marked reduction in bone formation and the development of an aplastic lesion without evidence of aluminum deposition in bone. Two patients, however, progressed from mild lesions of secondary hyperparathyroidism to overt osteitis fibrosa during the study.

In the 13 patients with initial bone biopsy lesions of osteitis fibrosa, serum iPTH levels were $818 \pm 201 \, \mu l$ Eq/ml at the start of treatment in those who failed to respond favorably to oral calcitriol therapy whereas values were $332 \pm 84 \, \mu l$ Eq/ml in patients who showed histologic improvement. Serum iPTH increased further to $1,275 \pm 297$, $p < 0.05$, in those who did not respond to treatment, and values at the end of the 12-month study period were substantially greater in these patients than in those with improved bone histology following oral calcitriol therapy, $315 \pm 130 \, \mu l$ Eq/ml, $p < 0.05$.

Discussion

The results of the current study indicate that secondary hyperparathyroidism remains a substantial problem in the clinical management of children undergoing CAPD/CCPD. Skelietal lesions of hyperparathyroidism were evident on initial evaluation in fully two-thirds of pediatric patients receiving long-term peritoneal dialysis, and more than half of the patients reevaluated after 1 year demonstrated progressive changes of osteitis fibrosa or failed to improve despite treatment with daily oral calcitriol. There were substantial differences, however, in the histologic and biochemical responses to oral calcitriol between patients with mild lesions and those with more advanced skeletal changes of secondary hyperparathyroidism. Patients with established osteitis fibrosa were more likely to show histologic progression and increases in serum iPTH levels than those with mild secondary hyperparathyroidism. Such findings suggest that the severity of secondary hyperparathyroidism prior to the start of treatment is an important determinant of the subsequent therapeutic response to oral calcitriol. This observation underscores the value of thorough, early diagnostic assessments of renal osteodystrophy in pediatric patients undergoing CAPD/CCPD and the potential importance of therapeutic interventions undertaken prior to the development of overt osteitis fibrosa [11].

The proportion of pediatric patients in the current study who had progressive changes of secondary hyperparathyroidism was considerably greater than expected This finding may be attributable, in part, to the severity of secondary hyperparathyroidism in many of the patients at entry into study. The mean serum iPTH level in patients with either mild lesions of secondary hyperparathyroidism or osteitis fibrosa was 501 ± 106, a value more than 50 times the upper limit of normal for the assay utilized in these studies. Such levels generally indicate the presence of severe secondary hyperparathyroidism in adult patients receiving regular hemodialysis [12].

Earlier reports indicated that adult hemodialysis patients with osteitis fibrosa generally improve following treatment with daily oral calcitriol as judged by reductions in the degree of tissue fibrosis, bone resorption surface and serum iPTH levels, results substantially different from the current observations in pediatric patients receiving peritoneal dialysis [1–3, 13]. It is now recognized, however, that secondary hyperparathyroidism may prove refractory to treatment with oral calcitriol in a number of adult patients undergoing long-term dialysis [14]. Thus, alternative approaches to daily oral calcitriol therapy have recently been evaluated in small numbers of adult patients undergoing regular hemodialysis; these include thrice weekly intravenous injections of calcitriol and large oral doses of calcitriol given once or twice per week [14–16]. Although the findings reported to date are encouraging, additional data will be required, including evaluations of the skeletal response to intermittent calcitriol administration, to adequately assess the efficacy of these newer approaches to treatment. Whether similar alternatives can be used with safety and efficacy in children with secondary hyperparathyroidism has yet to be determined.

It is of interest that recent preliminary data of Hercz et al. [17] indicate that the prevalence of the aplastic, or adynamic, lesion of renal osteodystrophy not associated with aluminum deposition in bone is quite high in adult patients receiving long-term peritoneal dialysis. Such findings are in striking contrast to those herein reported for pediatric patients undergoing CAPD/CCPD in whom aplastic lesions were found in only 5 of 59 initial bone biopsies. Patients with aplastic lesions of renal osteodystrophy often have normal or reduced serum levels of iPTH, a feature which suggests a state of relative hypoparathyroidism. Although the number of patients with aplastic lesions in the current series is inadequate to address this issue, it is possible that the more widespread use of calcium-containing medications as phosphate-binding agents in patients with chronic renal failure may result in oral calcium loading and markedly positive calcium balances in patients already exposed continuously to high levels of calcium in peritoneal dialysate. Thus, oral calcium supplementation and the concurrent use of active vitamin D sterols should be further evaluated as potential factors that contribute to the pathogenesis of aplastic lesions of bone that develop in the absence of bone aluminum deposition in patients with chronic renal failure.

Acknowledgements

Supported, in part, by USPHS grants DK-35423 and AR-35470 and by research funds of the Veterans Administration. Ms. Joanne Foley, RN, provided valuable assistance in the clinical evaluation and care of the patients involved in these studies, and Ms. Jeanenne O'Connor was responsible for the quantitative histologic studies of bone. Ms. Lisa Neuman provided secretarial assistance.

Summary

The course of renal bone disease in children undergoing regular peritoneal dialysis has not been fully evaluated. In particular, only limited data are available regarding the therapeutic efficacy of oral calcitriol in this group of patients. Thus, bone biopsies were done before and after 12 months of treatment with daily doses of oral calcitriol in 33 pediatric patients undergoing CAPD or CCPD. In 10/13 patients with initial biopsy findings of osteitis fibrosa, skeletal lesions either failed to improve or worsened after 12 months of treatment; in contrast, 4/6 patients with mild lesions of secondary hyperparathyroidism demonstrated histologic improvement. Serum PTH levels, determined using a mid-region assay, were higher in patients who failed to improve during oral calcitriol therapy, and values increased progressively in this group. Serum PTH levels were unchanged in those who improved. These data indicate that secondary hyperparathyroidism remains a substantial clinical problem in pediatric patients receiving regular peritoneal dialysis; treatment with daily doses of oral calcitriol is only partially effective for the control of this disorder. Further assessment of alternatives to daily oral calcitriol administration such as pulse oral or intermittent intraperitoneal calcitriol therapy is warranted in order to provide more effective approaches to the management of secondary hyperparathyroidism in pediatric dialysis patients.

References

1   Brickman AS, Sherrard DJ, Jowsey J, Singer FR, Baylink DJ, Maloney N, Massry SG, Norman AW, Coburn JW: 1,25-Dihydroxycholecalciferol: Effect on skeletal lesions and

plasma parathyroid hormone levels in uremic osteodystrophy. Arch Intern Med 1974;134:883–888.

2   Sherrard DJ, Coburn JW, Brickman AS, Singer FR, Maloney N: Skeletal response to treatment with 1,25-dihydroxyvitamin D in renal failure. Contrib Nephrol. Basel, Karger, 1980, vol 18, p 92.

3   Berl T, Berns AS, Huffer WE, Hammill K, Alfrey AC, Arnaud CD, Schrier RW: 1,25-Dihydroxycholecalciferol effects in chronic dialysis. A double-blind controlled study. Ann Intern Med 1978;88:774–780.

4   Gokal R, Ramos JM, Ellis HA, Parkinson I, Sweetman V, Dewar J, Ward MK, Kerr DNS: Histological renal osteodystrophy and 25–hydroxy-cholecalciferol and aluminum levels in patients on continuous ambulatory peritoneal dialysis. Kidney Int 1983;23:15–21.

5   Delmez JA, Fallon MD, Bergfeld MA, Gearing BK, Dougan CS, Teitelbaum SL: Continuous ambulatory peritoneal dialysis and bone. Kidney Int 1986;30:379–384.

6   Alfrey AC, Mishell MM, Burks J, Contiguglia SR, Rudolph H, Lewin E, Holmes JE: Syndrome of dyspraxia and multifocal seizures associated with chronic hemodialysis. Trans Am Soc Artif Intern Organs 1972;18:257–261.

7   Salusky IB, Fine RN, Kangarloo H, Gold R, Paunier L, Goodman WG, Brill JE, Gilli G, Slatopolsky E, Coburn JW: High dose calcitriol for control of renal osteodystrophy in children on CAPD. Kidney Int 1987;32:89–95.

8   Sedman AB, Miller NL, Warady BA, Lum GM, Alfrey AC: Aluminum loading in children with chronic renal failure. Kidney Int 1984;26:201–204.

9   Salusky IB, Coburn JW, Brill J, Foley J, Slatopolsky E, Fine RN, Goodman WG: Bone disease in pediatric patients undergoing dialysis with CAPD or CCPD. Kidney Int 1988;33:975–982.

10   Dixon WJ, Massey FJ: Introduction to Statistical Analysis. New York, McGraw-Hill, 1983.

11   Baker LRI, Abrams SML, Roe CJ, Faugere M-C, Fanti P, Subayti Y, Malluche HH: Early therapy of renal bone disease with calcitriol: A prospective double-blind study. Kidney Int 1989;36 (suppl 27):S140–S142.

12   Coburn JW, Slatopolsky E: Vitamin D, parathyroid hormone, and renal osteodystrophy; in Brenner B, Rector F, (eds): The Kidney. Philadelphia, Saunders, 1986, pp 1657–1729.

13   Healy M, Malluche HH, Goldstein DA, Singer FR, Massry SG: Effects of long-term therapy with 1,25 (OH)$_2$D$_3$ in patients with moderate renal failure (abstract). Arch Intern Med 1980;140:1030–1033.

14   Andress DL, Norris KC, Coburn JW, Slatopolsky EA, Sherrard DJ: Intravenous calcitriol in the treatment of refractory osteitis fibrosa of chronic renal failure. N Engl J Med 1989;321:274–279.

15   Slatopolsky E, Weerts C, Thielan J, Horst RL, Harter H, Martin KJ: Marked suppression of secondary hyperparathyroidism by intravenous administration of 1,25–dihydroxyc-holecalciferol in uremic patients. J Clin Invest 1984;74:2136–2143.

16   Fukagawa M, Okazaki R, Takano K, Kaname S, Ogata E, Kitaoka M, Harada S, Sekine N, Matsumoto T, Kurokawa K: Regression of parathyroid hyperplasia by calcitriol-pulse therapy in patients on long-term dialysis. N Engl J Med 1990;323:421–422.

17   Hercz G, Goodman WG, Pei Y, Segre GV, Coburn JW, Sherrard DJ: Low turnover bone disease without aluminum in dialysis patients (abstract) Kidney Int 1989;35:378.

William G. Goodman, MD, Nephrology Section (111R), Sepulveda VA Medical Center, 16111 Plummer Street, Sepulveda, CA 91343 (USA)

Morii H (ed): Calcium-Regulating Hormones. I. Role in Disease and Aging.
Contrib Nephrol. Basel, Karger, 1991, vol 90, pp 196–203

Clinical Trial of 26,26,26,27,27,27-Hexafluoro-1,25-Dihydroxyvitamin D₃ in Uremic Patients on Hemodialysis: Preliminary Report

Yoshiki Nishizawa[a], *Hirotoshi Morii*[a], *Yosuke Ogura*[b],
Hector F. DeLuca[c, 1]

[a]Second Department of Internal Medicine, Osaka City University Medical School,
Japan; [b]Kidney Center, Toranomon Hospital, Tokyo, Japan; [c]Department of
Biochemistry, University of Wisconsin, Madison, Wisc., USA

In normal man, the kidney is the only organ that produces significant quantities of $1,25(OH)_2D_3$. Thus, $1,25(OH)_2D_3$, the active form of vitamin D, is an effective treatment for renal osteodystrophy [1–3]. Other metabolites or derivatives of vitamin D, such as dihydrotachysterol (DHT) [4, 5], $25(OH)D_3$ [6, 7], $1\text{-}\alpha(OH)D_3$ [8, 9], and $24,25(OH)_2D_3$ [10] have been investigated and found to be clinically useful for uremic bone conditions. There is, therefore, much interest in the derivatives or analogues of vitamin D, which may be more potent and more effective in the treatment of renal osteodystrophy with fewer side effects.

A fluorinated analogue of vitamin D_3, 26,26,26,27,27,27-hexafluoro-$1,25(OH)_2D_3$ ($26,27\text{-}F6\text{-}1,25(OH)_2D_3$) was synthesized because it seemed likely that 24- and 26-hydroxylation of vitamin D_3 causes its inactivation during metabolism [11, 12]. This compound is more active than $1,25(OH)_2D_3$ in healing rickets, in increasing intestinal calcium transport

[1] On behalf of the Study Group for F6-1,25 D_3 for Uremic Patients, Japan. Participants: Masashi Suzuki, Hemodialysis Unit, Shrinrakuen Hospital, Niigata; Yoshindo Kawaguchi, Second Department of Internal Medicine, Tokyo Jikeikai University, School of Medicine; Yoshio Suzuki, Kidney Center, Toranomon Hospital, Tokyo; Seishi Inoue, Hemodialysis Unit, Hyogo College of Medicine, Nishinomiya; Masato Nishioka, Division of Internal Medicine, Sumiyoshigawa Hospital, Kobe; Yoshitaka Oda, Masao Akiyama, Division of Internal Medicine, Masuko Hospital, Kawaguchi; Ryo Syoji, Minami-Ohi Clinic, Tokyo; Noritsugu Imamura, Minami-Tamachi Clinic, Tokyo; Takashi Inoue, Tsutomu Tabata, Yutaka Furumitsu, Hemodialysis Unit, Inoue Hospital, Suita; Makoto Yamakawa, Hiroshi Nishitani, Hemodialysis Unit, Shirasagi Hospital, Osaka; Eiichi Chiba, Kiyotaka Omura, Fumie Noro, Kidney Center, Mikasa City General Hospital, Hokkaido.

and bone calcium mobilization [13], in causing the differentiation of HL-60 cells [14], and in inhibiting interleukin-2 production [15].

In this study, a clinical trial was done to evaluate the efficacy of 26,27-F6-1,25(OH)$_2$D$_3$ on the calcium and bone metabolism in uremic patients on hemodialysis.

Materials and Methods

Forty-three uremic patients on maintenance hemodialysis were the subjects of this study. There were 24 men and 19 women with a mean age of 50.9 ± 2.1 years. During selection, patients who were pregnant, who within the previous 3 months took drugs that influence calcium homeostasis, or who had serious complications, were excluded. The period of study was 14 weeks, including 2 weeks for observation before the trial began.

The initial dose of 26,27-F6-1,25(OH)$_2$D$_3$, administered orally was 0.05 µg/day for the first 2 weeks. Then the dose was increased by 0.05 µg every 2 weeks until a dose of 0.3 µg/day was reached or until serum calcium levels increased. When serum calcium levels were elevated near to 10.5 mg/dl, the dose was reduced to the preceding dose or the drug was discontinued.

26,27-F6-1,25(OH)$_2$D$_3$ was supplied by Sumitomo Pharmaceutical Co. Ltd., and Taisho Pharmaceutical Co. Ltd., Japan.

C-terminal PTH (c-PTH) was assayed with an INC-RIA kit, midportion PTH (m-PTH) with a Yamasa RIA kit, intact PTH by Allegro IRMA, and bone Gla protein (BGP) with a CIS kit.

The results are expressed as means $\pm$ SE. Student's t test was used to test the significance of differences.

Results

Background of the Patients

Serum levels of corrected calcium, phosphate, and magnesium were 8.79 ± 0.12, 5.47 ± 0.26 and 2.86 ± 0.08 mg/dl, respectively, before the administration of the drug. The baseline levels of c-PTH, m-PTH, intact PTH, and BGP were 4.80 ± 0.70, 17.8 ± 2.6, 165 ± 31 and 35.5 ± 5.8 ng/ml, respectively, showing that most of the subjects had secondary hyperparathyroidism. In some, this condition was advanced.

Daily Mean Dose

Serum calcium levels increased significantly at the mean dose of 0.08 µg/day. The levels of phosphate and magnesium did not change significantly at any dose. The serum level of PTH decreased significantly at the mean dose of 0.08 µg/day for intact PTH and at 0.14 µg/day for the other PTH measurements. The level of serum BGP increased significantly in week 12 of the study, at the mean dose of 0.18 µg/day, as did the level of alkaline phosphatase (table 1).

Table 1. The mean dose of F6-1,25(OH)$_2$D$_3$ and laboratory data

	Week 0	Week 2	Week 4	Week 6	Week 8	Week 10	Week 12
Mean dose, μg/day	–	0.05	0.056 ± 0.0026	0.79 ± 0.0047	0.116 ± 0.08	0.14 ± 0.01	0.175 ± 0.012
s-Ca, mg/dl (corrected)	8.79 ± 0.12	9.09 ± 0.14	9.00 ± 0.12	9.34 ± 0.12*	9.18 ± 0.12**	9.16 ± 0.14	9.09 ± 0.13*
s-P, mg/dl	5.47 ± 0.26	5.65 ± 0.21	5.57 ± 0.27	5.46 ± 0.21	5.36 ± 0.27	5.33 ± 0.27	4.99 ± 0.22
s-Mg, mg/dl	2.86 ± 0.08	2.80 ± 0.11	2.81 ± 0.08	2.83 ± 0.08	2.82 ± 0.08	2.76 ± 0.01	2.78 ± 0.07
BGP, ng/ml	35.5 ± 5.8	34.1 ± 4.4	35.0 ± 4.4	36.1 ± 4.0	38.6 ± 4.5	37.7 ± 4.8	40.5 ± 5.5*
AP[1]							
IU	198.0 ± 34.0	237.5 ± 71.8	182.3 ± 31.8	178.4 ± 46.9	177.5 ± 26.9	214.9 ± 55.0	222.1 ± 58.3*
KAU	10.68 ± 1.07	10.23 ± 1.07	11.05 ± 1.08	11.46 ± 1.17*	11.48 ± 1.15*	11.32 ± 1.16	11.65 ± 1.14*
C-PTH, ng/ml	4.80 ± 0.70	4.85 ± 0.91	4.90 ± 0.97	4.40 ± 0.88*	4.75 ± 0.95	4.18 ± 0.99**	4.67 ± 1.14
m-PTH, ng/ml	17.8 ± 2.6	17.8 ± 3.5	18.3 ± 3.4	17.2 ± 3.5	18.0 ± 3.4	15.4 ± 3.6**	15.9 ± 3.7
Intact-PTH, pg/ml	165 ± 31	170 ± 32	174 ± 30	148 ± 25	175 ± 37	122 ± 27*	125 ± 26**

The results are expressed as means ± SE, and paired t test was used for the significance of differences in comparing to the values at week 0; *p < 0.05; **p < 0.01.
[1] AP were measured in international units (IU) for 25 patients and in King–Armstrong units (KAU) for 18 patients.

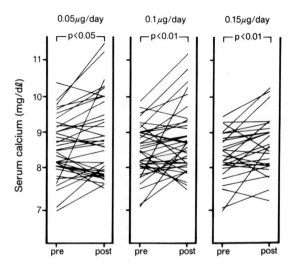

Fig. 1. Serum level of calcium significantly increased at the dose of 0.05 µg/day by dose comparison. Pre-calcium level before the trial: post-calcium level at each dose.

Daily Dose by Dose Comparison

To compare the efficacy of different doses, the results were collected for each dose in the different weeks of the study. By dose comparison, the dose of 0.5 µg/day was found to be effective in elevating serum calcium (fig. 1), 0.3 µg/day decreased the PTH level (fig. 2), and 0.25 µg/day increased the BGP level (fig. 3).

Discussion

The study group, which included three universities and nine centers for kidney disease and hemodialysis in Japan, investigated the effects of low doses of 26,27-F6-1,25(OH)$_2$D$_3$ in uremic patients on hemodialysis. We found that this analogue significantly increased serum calcium at the very low dose of 0.05–0.08 µg. The dose of 0.125–0.25 µg of 1,25(OH)$_2$D$_3$ seems to be the minium dose for calcium mobilization, so 26,27-F6-1,25(OH)$_2$D$_3$ is 2.5–5 times as potent as 1,25(OH)$_2$D$_3$. To 1 patient not in this study with idiopathic hypoparathyroidism and normal renal function, we first administered 26,27-F6-1,25(OH)$_2$D$_3$ and then switched to 1,25(OH)$_2$D$_3$. The urinary calcium/creatinine ratio increased even at the

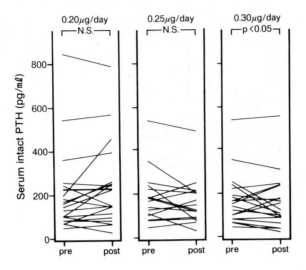

Fig. 2. Serum intact PTH significantly decreased at the dose of 0.30 μg/day by dose comparison.

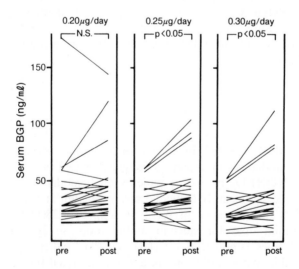

Fig. 3. Serum bone Gla-protein (BGP) significantly increased at the dose of 0.25 μg/day by dose comparison.

daily dose of 0.02 μg of 26,27-F6-1,25(OH)$_2$D$_3$. Serum ionized calcium increased and the hypocalcemic symptoms improved when the dose of the analogue was increased to 0.8 μg and then to 1.2 μg. For this patient the same doses of 1,25(OH)$_2$D$_3$ failed to improve the hypocalcemic symptoms or to increase the level of ionized calcium. The results with that subject suggested that about 2.0–2.5 μg of 1,25(OH)$_2$D$_3$ has the same effect as about 0.8–1.2 μg of 26,27-F6-1,25(OH)$_2$D$_3$. In other words, in that case, the fluorinated analogue had a potency 2 to 3 times that of 1,25(OH)$_2$D$_3$. The results of our study of uremic patients were in agreement with this finding.

This analogue is more potent in terms of various actions of vitamin D. Tanaka et al. [13] showed that fluorinated vitamin D is about 5 times as effective than 1,25(OH)$_2$D$_3$ in healing rickets and about 10 times more effective in increasing intestinal calcium transport and calcium mobilization of rats deficient in vitamin D and fed on a low-calcium diet. Kiriyama et al. [16] reported that the biological activity of 26,27-F6-1,25(OH)$_2$D$_3$ in chicks deficient in vitamin D is greater than that of 1,25(OH)$_2$D$_3$ and that it is about four times as potent in mobilizing bone calcium. In our own studies we have found that 26,27-F6-1,25(OH)$_2$D$_3$ is 5 times more active than 1,25(OH)$_2$D$_3$ in calcium mobilizing bone calcium in both parathyroidectomized rats or 5/6-nephrectomized rats. Inaba et al. [14] showed that 26,27-F6-1,25(OH)$_2$D$_3$ is 10 times more active than 1,25(OH)$_2$D$_3$ in inhibiting the proliferation or inducing the differentiation of HL-60 cells. Yukioka et al. [15] found 26,27-F6-1,25(OH)$_2$D$_3$ to be ten times more potent than 1,25(OH)$_2$D$_3$ in inhibiting interleukin-2 production with PHA. These results show that 26,27-F6-1,25(OH)$_2$D$_3$ is more potent than 1,25(OH)$_2$D$_3$ *in vitro*, *in vivo* in animals, and in patients reported here.

The explanation of the high potency of this compound is not known. It has a somewhat lower affinity for the 1,25(OH)$_2$D$_3$ receptor from chick intestine [13–17] or in HL-60 cells [14], and it has a similar biological half-life [18]. Perhaps a metabolite such as 26,27-F6-1,23,25(OH)$_2$D$_3$ may accumulate in target tissue, resulting in a more prolonged effect, or perhaps the 26,27-F6-1,25(OH)$_2$D$_3$-receptor complex is bound more tightly to DNA than the 1,25(OH)$_2$D$_3$-receptor complex [17].

Summary

A clinical trial was done by the Group, Japan to evaluate the efficacy of 26,27-F6-1,25(OH)$_2$D$_3$ on the calcium and bone metabolism of 43 uremic patients on hemodialysis, 24 men and 19 women with a mean age of 50.9 ± 2.1 years. The initial dose administered orally was 0.05 μg/day for 2 weeks. Then the dose was increased every 2 weeks by 0.05 μg each time until the dose of 0.3 μg/day was reached or until serum calcium increased. 26,27-F6(OH)$_2$D$_3$ increased serum calcium levels significantly at a mean dose of 0.08 ± 0.03 μg/day and at 0.05 μg/day of dose comparison in hemodialyzed patients. It decreased the serum level of PTH significantly at a mean dose of 0.14 ± 0.06 μg/day and at 0.3 μg/day by dose comparison. The serum level of bone Gla protein increased singificantly at a mean dose of 0.18 ± 0.07 μg/day and at 0.25 μg/day by dose comparison in the same patients. These results suggest that 26,27-F6-1,25(OH)$_2$D$_3$ has a higher potency in calcium mobilization than 1,25(OH)$_2$D$_3$ in uremic patients on hemodialysis.

References

1 Coburn JW, Hartenbower DL, Brickman AS: Advances in vitamin D metabolism as they pertain to chronic renal disease. Am J Clin Nutr 1976;29:1283–1299.

2 Healy MD, Malluche HH, Goldstein DA, Singer FR, Massry SG: Effects of long-term therapy with calcitriol in patients with moderate renal failure. Arch Intern Med 1980;140:1030;–1033.

3 Brickman AS, Coburn JW, Massry SG, Norman AW: 1,25-Dihydroxy-vitamin D$_3$ in normal man and patients with renal failure. Ann Intern Med 1974;80:161–168.

4 Kaye M, Chatterji G, Cohen EF, Sagar S: Arrest of hyperparathyroid bone disease with dihydrotachysterol in patients undergoing chronic haemodialysis. Ann Intern Med 1970;73:225–233.

5 Cordy PE, Mills DM: The early detection and treatment of renal osteodystrophy. Miner Electrolyte Metab 1981;5:311–320.

6 Teitelbaum SL, Bone JM, Stein PM, Gildew JJ, Bates M, Boisseau VC, Avioli LV: Calcifediol in chronic renal insufficiency: Skeletal response. JAMA 1976;235:164–167.

7 Recker R, Schoenfeld P, Letteri J, Slatopolsky E, Goldsmith R, Brickman A: The efficacy of calcifediol in renal osteodystrophy. Arch Intern Med 1978;138:857–863.

8 Madsen S, Øgaard K: 1-Alpha-hydroxycholecalciferol treatment of adults with chronic renal failure. Acta Med Scand 1976;200:1–5.

9 Peacock M: The clinical uses of 1α-hydroxyvitamin D$_3$. Clin Endocrinol 1977;7:1–246.

10 Llach F, Brickman AS, Singer FR, Coburn JW: 24,25-Dihydroxycholecalciferol, a vitamin D sterol with qualitatively unique effects in uremic man. Metab Bone Dis Rel Res 1979;2:11–16.

11 Kobayashi Y, Taguchi T, Mitsuhashi S, Eguchi T, Ohshima E, Ikegami N: Studies on organic fluorine compounds. XXXIX. Studies on steroids. LXXIX. Synthesis of 1α,25-dihydroxy-26,26,26,27,27,27-hexafluorovitamin D$_3$. Chem Pharm Bull (Tokyo) 1982;30:4296–4303.

12 Stern PH, Mavreas T, Tanaka Y, DeLuca HF, Ikekawa N, Kobayashi Y: Fluoride substitution of vitamin D analogs at C-26 and C-27: Enhancement of activity of 25-hydroxyvitamin D but not of 1,25-dihydroxyvitamin D on bone and intestine in vitro. J Pharm Exp Ther 1984;229:9–13.

13 Tanaka Y, DeLuca HF, Kobayashi Y, Ikekawa N: 26,26,26,27,27,27-Hexafluoro-1,25-dihydroxyvitamin D$_3$: A highly potent, long-lasting analog of 1,25-dihydroxyvitamin D$_3$. Arch Biochem Biophys 1984;229:348–354.

14 Inaba M, Okuno S, Nishizawa Y, Yukioka K, Otani S, Matsui-Yuase I, Morisawa S, DeLuca HF, Morii H: Biological activity of fluorinated vitamin D analogs at C-26 and C-27 on human promyelocytic leukemia cells, HL-60. Arch Biochem Biophys 1987;258:421–425.

15   Yukioka K, Otani S, Matsui-Yuasa I, Goto H, Morisawa S, Okuno S, Inaba M, Nishizawa Y, Morii H: Biological activity of 26,26,26,27,27,27-hexafluorinated analogs of vitamin D3 in inhibiting interleukin-2 production by peripheral blood mononuclear cells stimulated phytohemagglutinin. Arch Biochem Biophys 1988;260:45–50.

16   Kiriyama T, Okamoto S, Suzuki H, Nagata A, Izumi M, Moriki H, Nagataki S: Biological activity of 26,26,26,27,27,27-hexafluoro-1,25-dihydroxyvitamin D_3 in the chick. Acta Endocrinol 1989;121:520–524.

17   Inaba M, Okuno S, Inoue A, Nishizawa Y, Morii H, DeLuca HF: DNA binding property of vitamin D_3 receptors associated with 26,26,26,27,27,27-hexafluoro-1,25-dihydroxyvitamin D_3. Arch Biochem Biophys 1989;268:35–39.

18   Nagata A, Hamma N, Katsumata T, Sato R, Komuro R, Iba K, Yoshitake A, Kiriyama T, Okamoto S, Nagataki S, Morii H: Mechanism of action of 26,27-hexafluoro-1,25($OH)_2D_3$ (abstract). 7th Workshop on Vitamin D, 1988, p 27.

Yoshiki Nishizawa, Assoc. Prof., Second Department of Internal Medicine, Osaka City University Medical School, 1-5-7 Asahi-machi, Abeno-ku, Osaka 545 (Japan)

Calcium, Diabetes mellitus and Aging

Morii H (ed): Calcium-Regulating Hormones. I. Role in Disease and Aging.
Contrib Nephrol. Basel, Karger, 1991, vol 90, pp 206–211

Calcium, Parathyroids and Aging

Takuo Fujita

Third Division, Department of Medicine, Kobe University School of Medicine,
Kobe, Japan

Since life was probably created in seawater with abundant calcium content on earth which might be called a planet of calcium, calcium is indispensable in all the functions of each cell. When some form of life decided to come out of the abundance of calcium to live on land, inevitable calcium deficiency occurred, with aggravation in aging. Parthyroid glands secrete PTH which mobilizes calcium from bone to restore serum calcium which may decrease mildly and transiently in calcium deficiency. PTH may therefore be called a hormone for calcium deficiency. It may not be a coincidence that no parathyroid glands are found in fish, which are living in an environment with abundant calcium. Calcium, parathyroids and aging thus appear to be three imporant factors interrelated to each other to control the body and cell function.

Role of Calcium in Physiology and Nutrition

Calcium is the 5th most abundant element in the human body and is unique in its distribution. While 99% of calcium in the body is distributed in the skeleton to provide the hardness it requires to support the body, a small amount circulating in the blood serves to maintain a constant serum calcium level, which is the most strictly regulated biological constant [1]. Serum calcium should be maintained constant because it is vital for the cells to maintain a constant level of intracellular free cytosolic calcium, which is only 1/10,000 that of extracellular calcium level. Such a vast intra-extracellular concentration gradient is unique for calcium and essential for the maintenance of cell membrane integrity and all cell functions including secretion, excitation, locomotion differentation, and proliferation. Signal transduction heavily depends on this vast extra- and intracellular calcium concentration gradient. Whenever such a concentration gradient is

blunted between the extra- and intracellular compartment, serious disturbances occur in the signal transduction.

Parathyroid hormone is known to increase intracellular calcium in at
least 8 kinds of cells, osteoblasts, renal tubular cells, lymphocytes, neu-
trophilic leucocytes, red blood cells, myocardial cells, vascular smooth
muscle cells and pancreatic β-cells [2, 3]. Since calcium deficiency stimulates PTH secretion, calcium deficiency tends to increase intracellular
calcium, blunting the extra- and intracellular calcium concentration gradient. A shift of calcium from the skeleton to the soft tissue like blood vessel
and from the extracellular to the intracellular compartment thus takes
place on calcium deficiency through the action of PTH. Since aging is
characterized by calcium deficiency, such progressive increase of soft tissue
and intracellular calcium leading to a blunting of calcium concentration
gradients between compartments invariably occurs in aging. Aging is also
associated with decreasing appetite and decrease of gastrointestinal absorptive function leading to less calcium intake and absorption. Age-bound
decrease of renal function with a fall of 1,25(OH) vitamin D synthesis also
decreases intestinal calcium absorption. All these factors contribute to
calcium deficiency in aging. Aging may also be regarded as a slowly
progressive renal insufficiency.

Consequences of the Rise of Intracellular Calcium Concentration and Blunting of Intra- and Extracellular Calcium Concentration

All cell death is characterized by an increase of intracellular calcium as
exemplified by hepatic cell necrosis on exposure to carbon tetrachloride,
progressive muscular dystrophy due to hereditary membrane abnormality,
myocardial and nerve cells on ischemia and cells infected by viruses or
bacteria [4]. Increase of cytoplasmic free calcium may therefore be called
'the final common path' of cell disease and cell death. Aging as a background of diseases is also characterized by an increase of intracellular
calcium. Diseases typically associated with aging include hypertension,
arteriosclerosis, diabetes mellitus and dementia.

Hypertension is caused by contraction of the vascular smooth muscle
induced by the increase of intracellular calcium. Although the acute
action of PTH is hypotensive, mediated by relaxation of the smooth
muscle, chronic parathyroid hyperfunction in response to calcium deficiency invariably causes an increase in intracellular calcium in vascular
smooth muscle and hypertension in humans and animals with experimental
hypertension, including spontaneously hypertensive rats [5]. Disturbance of
calcium transport due to generalized membrane abnormalities manifested

by increased urinary calcium secretion and decreased intestinal calcium absorption leads to calcium deficiency which is aggravated by decreased calcium intake. Oral calcium supplement understandably had a favorable influence on hypertension. Hypertension is thus an example of calcium deficiency.

Diabetes mellitus is another example of calcium deficiency. Calcium deficiency leads to PTH hypersection and increase of the intracellular calcium of pancreatic β-cells. Ensuing blunting of intra- extracellular calcium concentration gradient interferes with the signal transduction mechanism of insulin secretion by β-cells causing a derangement of timely insulin secretion in response to glucose load. Calcium supplement would cause recovery from such a disturbance of insulin secretion. In many diabetics, the calcium intake is not sufficient to meet the requirement of the body, because adequate calcium intake has not been one of the goals in the dietary therapy of diabetes mellitus. Such persistent calcium deficiency may not only produce osteopenia, but also blunting of the intra- extracellular compartment to aggravate cell injury in diabetic complications such as nephropathy, neuropathy and retinopathy.

Calcium Preparation in Intervention Studies of Calcium Supplementation

Calcium supplement has so far been given mainly as calcium carbonate. Oyster shell electrolysate (OSE) was recently developed through subjecting powdered oyster shell to electrolysis at a high temperature. This preparation has a characteristic lamellar appearance under the scanning electron microscope which is to be distinguished from oyster shell powder, calcium carbonate and calcium oxide [6]. This material is also called active absorbable calcium (AACa) because of its high efficiency in intestinal absorption even in the absence of vitamin D. Complete balance studies in 4 healthy elderly human subjects indicated a much better bioavailability of AACa than that of calcium carbonate calcium lactate.

Short-term absorption studies in young normal subjects revealed no significant difference in the rise of serum calcium during the first 4 h after the oral administration of AACa and calcium carbonate, but serum phosphorus was significantly lower after administration of AACa than $CaCO_3$, indicating a higher phosphate-binding capacity of AACa, which would represent a promising agent as a phosphate binder in chronic renal failure [7].

Hypertension is in a group of hospitalized elderly patients with hypertension, blood pressure was constantly monitored and the effect of AACa

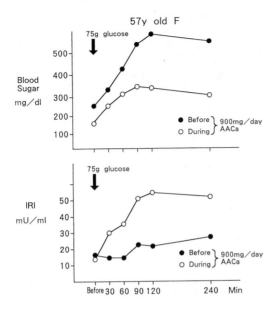

Fig. 1. Effect of oyster shell electrolysate (AACa) on oral glucose tolerance test in steroid-induced diabetes mellitus.

on blood pressure was tested against placebo and calcium carbonate in a crossover design.

Blood pressure fell significantly during the use of AACa compared to placebo or CaCO₃. Serum ionized calcium was also higher and PTH lower when using AACa than with placebo or calcium carbonate. AACa was more efficiently absorbed than CaCO₃ and apparently suppressed PTH better to normalize the blunted inter- extracellular calcium concentration gradient.

Similar results were obtained in diabetes mellitus. Oral administration of 3 g of calcium lactate was shown to have a favorable influence on insulin secretion in diabetes mellitus [8]. On the use of AACa supplying 900 mg elementary calcium, the results of 75 g glucose loading was definitely more favorable in a 57-year-old female diabetic (fig. 1). Blood sugar was lower, and blood insulin higher. In another 61-year-old female diabetic, the daily blood sugar profile definitely improved in response to 900 mg calcium supplement in the form of AACa (fig. 2). In a 65-year-old male diabetic, such an improvement in the control of diabetes mellitus in response to calcium supplement in the form of AACa was extended over a 24-hour period (fig. 3). Significantly more insulin was secreted and a lower blood sugar level was achieved in response to calcium supplement. Special care

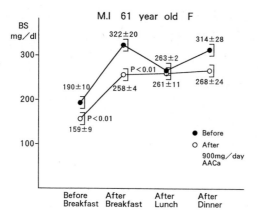

Fig. 2. Effect of oyster shell electrolysate (AACa) on blood sugar profile in diabetes mellitus.

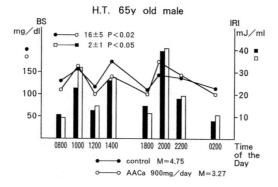

Fig. 3. Effect of oyster shell electrolysate (AACa) on blood sugar and insulin profile in diabetes mellitus.

should be taken to use the calcium source of high bioavailability for the assessment of the effect of calcium supplementation. Calcium carbonate is not necessarily the best agent available for this purpose.

Summary

Calcium is unique in its distribution in living organisms with an extremely high hard and soft tissue and extra- intracellular concentration gradient. Calcium deficiency through stimulating parathyroid hormone secretion tends to blunt such a difference by paradoxically increasing the calcium concentration in the soft tissue and intracellular compartment. Since aging is associated with the progressive aggravation of calcium deficiency, such blunting also progresses with aging. The dysfunction, damage and death of cells occurring in all diseases is always associated with a blunting of the extra- and intracellular calcium components. Calcium supplement especially with highly biologically available active absorbable calcium, was associated with the suppression of parathyroid hormone secretion and the normalization of a such blunting of intercompartmental distribution of calcium examples in hypertension and diabetes mellitus with evident improvement of clinical manifestations and laboratory tests.

References

1   Fujita T: Aging and calcium. Miner Electrolyte Metab 1986;12:149–156.
2   Fujii Y, Fukase M, Tsutsumi M, Miyauchi A, Tsunenari T, Fujita T: Parathyroid hormone control of free cytosolic calcium in the kidney. J Bone Mineral Res 1988;3:525–532.
3   Yamada H, Tsutsumi M, Fukase M, Fujimori M, Yamamoto Y, Miyauchi A, Fujii Y, Nada T, Fujii Y, Fujita T: Effects of human PTH-related peptide and human PTH on cyclic AMP production and cytosolic free calcium in an osteoblastic cell clone. Bone Miner 1989;6:45–54.
4   Faber JL: The role of calcium in cell death. Life Sci 1981;29:1289–1295.
5   Kazda S, Garthoft B, Luckhaus G: Calcium and malignant hypertension in animal experiment. Effects of experimental manipulation of calcium influx. J Nephrol 1986;6(suppl 1):145–150.
6   Fujita T, Fukase M, Nakada M, Koishi M: Intestinal absorption of oyster shell electrolysate. Bone Miner 1988;4:321–327.
7   Fujii Y, Tsutsumi M, Shimazu K, Negishi H, Fujita T: Active absorbable calcuim as a phosphate binders in dialysis patients. J Bone Miner Metab 1990;8:26–29.
8   Fujita T, Sakagami Y, Tomita T, Okamoto Y, Oku H: Insulin secretion after oral calcium load. Endocrinol Jpn 1978;25:645–648.

Takuo Fujita, MD, 3rd Department of Internal Medicine, Kobe University School of Medicine, 5-1 Kusunoki-cho, 7 chome, Chuo-ku, Kobe, Hyogo 650 (Japan)

Morii H (ed): Calcium-Regulating Hormones. I. Role in Disease and Aging.
Contrib Nephrol. Basel, Karger, 1991, vol 90, pp 212–216

Calcium Metabolism and Osteopathy in Diabetes mellitus

An-Hua Shao, Fu-Gi Wang, Yuan-Feng Hu, Li-Ming Zhang

Department of Internal Medicine, Department of Nuclear Medicine, Shanghai First People's Hospital, Shanghai, People's Republic of China

There is considerable evidence that many patients with diabetes of either insulin-dependent (IDDM) or non-insulin-dependent diabetes mellitus (NIDDM) may have a moderate-degree reduction of bone mineral content (BMC) [1]. In this study, we observed calcium metabolism and osteopathy in 11 IDDM patients on insulin and 19 NIDDM patients treated with an oral hypoglycemic agent, primarily the sulfonylureas.

Materials and Methods

Subjects

Thirty diabetics with normal kidney function, 11 male and 19 female, mean age 53.6 ± 14 years (23–70). Of these 11 were IDDM, 4 male and 7 female, mean age 38.6 ± 6.3 years (22–52), mean duration of diabetes 8.54 ± 4.1 years, and 19 NIDDM patients, 7 male and 12 female, mean age 62.5 ± 6.5 years, mean duration of diabetes 10.2 ± 7.1 years, and matched for age and sex with those normal subjects (table 1). None had other disease or received drugs known to interfere with calcium and bone metabolism.

Laboratory Methods

Blood samples were drawn in the early morning after an overnight fast. Blood glucose was measured by the glucose oxidation technique, HbA1 was estimated by the microcolumn method. Serum calcium (Ca) was determined by the EDTA method. Serum phosphate (P) by colorimetry. Alkaline phosphatase (AKP) was measured by the KA method. Serum parathyroid hormone (PTH) and calcitonin (CT) were estimated by radioimmunoassay. Lumbar spine density 1–5 was determined by roentgenogram.

Results

The study showed that the mean levels of serum Ca, P, and CT in diabetics of either IDDM or NIDDM patients were not significantly different as compared with those of controls. There was a significant

Table 1. Characteristics of diabetics and controls

	Controls	Diabetics	IDDM	NIDDM
Sex				
Male	11	11	4	7
Female	19	19	7	12
Mean age, years	50.5 ± 14.7	53.6 ± 14.0	38.6 ± 6.3	62.5 ± 6.5
	(22–70)	(23–70)	(22–52)	(50–70)
Mean duration of diabetes, years	–	9.6 ± 6.2	8.54 ± 4.1	10.2 ± 7.1

Table 2. Laboratory test in 30 diabetes mellitus patients and 30 controls ($\bar{x} \pm SD$)

	Controls (n = 30)	Diabetes mellitus (n = 30)	IDDM (n = 11)	NIDDM (n = 19)
FBG, mmol/l	4.62 ± 0.65	9.40 ± 2.78***	9.44 ± 2.50***	9.10 ± 2.70**
HbA1, %	6.03 ± 0.82	8.82 ± 3.29**	9.13 ± 4.17**	8.77 ± 2.64*
Ca, mmol/l	2.31 ± 8.00	2.30 ± 0.19	2.37 ± 0.28	2.27 ± 0.13
P, mmol/l	1.90 ± 0.31	1.84 ± 0.48	1.92 ± 0.44	1.67 ± 0.58
Cr, mmol/l	93.32 ± 24.77	100.29 ± 57.71	86.37 ± 11.50	105.50 ± 54.00
AKP, KA U	5.40 ± 1.47	6.85 ± 3.27*	8.50 ± 4.44*	6.11 ± 2.37
PTH, ng/l	0.87 ± 0.28	1.24 ± 0.61**	1.36 ± 0.50**	1.11 ± 0.64
CT, pg/l	54	53.82 ± 2.43	54.57 ± 1.90	53.39 ± 2.63

* $p < 0.05$;** $p < 0.01$;*** $p < 0.001$.

Table 3. Effect of age and sex on osteoporosis in diabetes mellitus

Age years	Male n	Female n	Total n
61–70	2	3	5
51–60		5	5
41–50			
31–40		1	1
21–30			

difference in the increase of mean values of PTH and AKP in IDDM patients ($p < 0.05$), but not in those of NIDDM (table 2).

Osteoporosis was shown by X-ray film in 11 of 21 diabetics vs. 6 of 21 controls. Between the two diabetic groups; 10 of 15 had osteoporosis which occurred in the NIDDM group, mainly in elderly women (table 3).

Table 4. Incidence of osteoporosis in diabetics and controls

Degree of osteoporosis	Controls		IDDM		NIDDM	
	male, n	female, n	male, n	female, n	male, n	female, n
Normal	3	12	2	3	1	4
Slight[1]	2	4		1	2	4
Severe[2]						4

[1] Less trabecular or thinner cortical of the bone.
[2] Compression fracture or double-concave deformation of the spine.

Among them, 4 cases showed bone lesions more seriously involved on roentgenogram, 2 had compression fracture and the other 2 showed double-concave deformation of the spine. Osteoporosis was present in 1 of the 6 IDDM patients only (table 4).

Discussion

Our data indicated that serum PTH was significantly increased in the IDDM group. Factors such as age, renal function, duration of diabetes and clinical control must be considered. Forero et al. [2] have shown that serum PTH is increased with advanced age or with declined renal function. However, the IDDM group patients were younger and with normal renal function. Albright and Reifenstein [3] showed that osteoporosis occurred in long-standing and poorly controlled diabetes, and many investigators [4–6] suggested that poorly controlled diabetics had massive glycosuria with calcium salt excretion as well as with intestinal calcium malabsorption due to lowered vitamin D levels which would lead to negative calcium balance and secondary increase of PTH secretion. In the present study, plasma HbA1 was taken as the index for diabetic control. However, there was no significant difference in plasma HbA1 between the IDDM and NIDDM groups. Duration of diabetes in the IDDM group was shorter than that in the NIDDM group. Osteoporosis can be caused by increased PTH secretion which may accelerate resorption of bone. In the IDDM patients, the incidence of osteoporosis was lower than that in the NIDDM group. This might probably be due to the lesser sensitivity of the roengenogram in reflecting bone changes; only in those showing a more than 30% loss of bone mineral content does the roentgenogram become conspicuous and the diagnosis of osteoporosis established [7]. Although the mean PTH values were not significantly increased in NIDDM patients, the incidence of osteoporosis was higher and most of them were elderly females. This is related to diabetes, advanced age and estrogenic hormone deficiency. Auwer et al. [8] used single and dual photon absorptiometry to examine the bone mineral content (BMC) of 31 diabetics, they found that the BMC of the lumbar spine was significantly reduced in female diabetics, but not in males. Isaia et al. [9] reported that the lumbar BMC was detected by double photon absorptiometry in 40 female NIDDM patients. The BMC of the lumbar spine was significantly lower in these female diabetics. Levin et al. [1] measured the bone mass of the forearm of diabetics by the photoabsorption technique and compared the results of three different treatment groups, i.e. diet alone, on insulin, or on oral hypoglycemic agents. A significantly greater loss of bone mass was seen in patients taking oral hypoglycemic agents than those treated by diet alone or on insulin. Sulfonylurea would increase the level of cyclic AMP by interfering with its phosphodiesterase-catalytic degradation. The cyclic AMP would accelerate bone resorption and lead to osteopenia or osteoporosis. The NIDDM patients were taking primarily sulfonylurea for a long period. Thereby, it might by one of the contributing factors causing higher incidence of osteoporosis in this group. Many factors such as age, sex, body weight, medical treatment, life style, social and economic status, etc., can influence the process of osteoporosis. Hence, the mechanism of diabetic osteoporosis is also much more complicated and so far unelucidated. There is still much debate [10].

Summary

To assess the changes of calcium metabolism and osteopathy in patients with diabetes. Serum Ca, P, AKP, PTH, CT, plasma fasting blood glucose (FBG) and HbA1 as well as X-ray film of the lumbar spine were measured in 30 diabetes patients; 11 were IDDM and 19 were NIDDM as compared to controls matched for age and sex. There were no significant differences in Ca, P, and CT values in serum between the IDDM and NIDDM patients and controls, whereas the serum levels of PTH and AKP were significant increased in IDDM patients. The incidence of osteoporosis which was shown by X-ray film in NIDDM patients was higher than in those of controls. No correlation between PTH value and osteoporosis or clinical control of diabetes was observed.

References

1   Levin ME, Boisseau VC, Avioli LV: Effects of diabetes mellitus on bone mass in juvenile and adult-onset diabetes. N Engl J Med 1976;294:241–245.
2   Forero MS, Klein RF, Nissenson RA, Heath H, Neison K, Arnaud D, Riggs BL: Effect

of age on circulating immunoreactive and bioactive parathyroid hormone levels in women. J Bone Mineral Res 1987;2:363–366.

3   Albright F, Reifenstein EC: Parathyroid glands and metabolic bone disease. Baltimore, Williams & Wilkins, 1948, pp 1–150.

4   Nair PM, Madshad S, Christensen C, Faber OK, Binder C, Transbol I: Bone mineral loss in insulin-treated diabetes mellitus. Acta Endocrinol 1979;90:463–472.

5   Schedl HP, Heath H, Wenger J: Serum calcitonin and parathyroid hormone in experimental diabetes. Endocrinology 1978;103:1368–1371.

6   Saito K: Bone change in diabetes mellitus (abstract). Nippon-Seikelgeka-Gakkai-Zasshi 1989;62:1189–1190.

7   Edeikeu J, Hodes PJ: Roentgen diagnosis of disease of bone, ed 2. Baltimore, Williams & Wilkins, 1973, vol 1, pp 1–407.

8   Auwerx J, Dequeker J, Bouillon R, Geusens P, Nijs J: Mineral metabolism and bone mass at peripheral and axial skelecton in diabetes mellitus. Diabetes 1988;37:8–12.

9   Isaia G, Bodrato L, Carlevatto V, Mussetto H, Salamanu G, Molinatti GM: Osteoporosis in type II diabetes (abstract). Acta Diabetol Lat 1987;24:305.

10 Heath H, Melton J, Chu CP: Diabetes mellitus risk of skeletal fracture. N Engl J Med 1980;303:567–570.

Dr. An-Hua Shao, Department of Internal Medicine, Shanghai First People's Hospital, 190 North Su Zhou Road, Shanghai, 200085 (People's Republic of China)

Morii H (ed): Calcium-Regulating Hormones. I. Role in Disease and Aging.
Contrib Nephrol. Basel, Karger, 1991, vol 90, pp 217–222

Parathyroid Hormone Secretion in Diabetes mellitus

Takahiko Kawagishi, Hirotoshi Morii, Kiyoshi Nakatsuka,
Kyoko Sasao, Koichi Kawasaki, Takami Miki, Yoshiki Nishizawa

Second Department of Internal Medicine, Osaka City University Hospital,
Osaka, Japan

It is now well established that there is a reduced incidence of hyper-parathyroidism in diabetic patients under hemodialysis [1, 2]. Avram et al. [3] were the first to show that diabetic patients had significantly lower parathyroid hormone levels than nondiabetic patients before the first dialysis. Morii et al. [4] reported that the incidence of bone loss showed less changes in diabetes mellitus under hemodialysis (HD/DM) than in nondia-betics under hemodialysis (HD/non-DM) which could be due to the low PTH level. Kikunami et al. [5] demonstrated that secondary hyperparathy-roidism in HD/DM developed slower than in HD/non-DM. On the other hand, it was pointed out that mean bone mass at the peripheral skelton is significantly decreased in diabetic patients with normal kidney function who had diminished serum calcium and phosphorus concentrations but normal PTH and active vitamin D levels [6]. So it is a problem of interest how PTH is secreted to stimuli before onset of nephropathy in diabetes mellitus. The present investigation was aimed to study whether the alterna-tions in parathyroid function in DM without renal dysfunction was found or not.

Methods

We studied 6 diabetic men (mean age, 62 years; range 57–67) and 6 aged-matched women (mean age, 66 years; range 61–69). None of the subjects had renal failure (serum creatinine <1.5 mg/dl) or took calcium, vitamin D or drugs affecting calcium metabolism. Each subject received a total of 2.0 g of oral phosphate daily as 2 divided doses on 5 consecutive days. Phosphate was a mixture of 14.8 g Na_2HPO_4 and 3.2 g NaH_2PO_4. Each took sodium phosphate at 8 a.m. and 8 p.m. The phosphate was tolerated well by all subjects. After the subjects were fasted overnight, blood and urine samples were obtained daily before and 2 h after the morning administration of phosphate. Blood concentrations of ionized calcium were measured with an electrode method and serum phosphorus was measured by

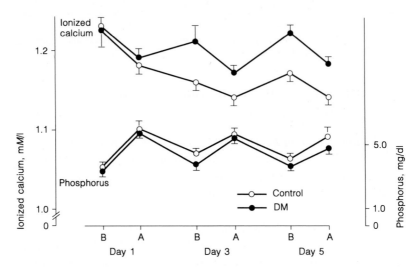

Fig. 1. Response of serum phosphorus and blood ionized calcium to phosphate administration. Values are means ± SEM for all subjects before (B) and 2 h after (A) phosphate
administration.

enzyme assay. Intact parathyroid hormone (PTH) concentrations were determined with the
sandwich method utilizing an amino- and carboxyl-terminal specific immunoradiometric
assay. All results are expressed as means ± SEM at each time point. The data were analyzed
by the paired or unpaired Student's t test and analysis of variance.

Results

The rise in serum phosphorus levels after phosphate administration
was similar in both diabetic and nondiabetic groups (fig. 1). The increase
became significant within 2 h of phosphate administration on day 1.
Phosphorus concentrations remained above the baseline throughout the
study period and continued to increase daily after phosphate administration. Blood ionized calcium levels were reduced from 1.23 ± 0.02 to 1.17 ±
0.01 mM/l in the diabetic group and from 1.23 ± 0.01 to 1.14 ± 0.01 mM/l
in the control group. Concomitant with the early fall in blood ionized
calcium levels, the intact PTH levels rose in the control group 2 h after the
first dose of phosphate was administered (fig. 2). In the control group, the
intact PTH levels rose daily and remained elevated throughout the study,
while in the diabetic group intact PTH did not respond significantly. In the
diabetic group, the intact PTH levels increased to 56% above the baseline
but remained within the normal range. In the control group, the intact

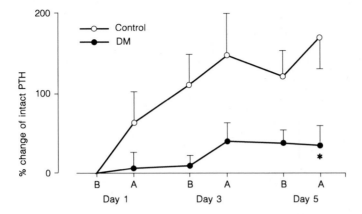

Fig. 2. Response of intact PTH to phosphate administration. Values are means $\pm$ SEM for all subjects before (B) and 2 h after (A) phosphate administration and denote changes from baseline. * $p < 0.05$ vs. control.

PTH levels increased almost 2.6-fold above the baseline by day 5 (fig. 2). The levels in the control group rose well above the upper limits of normal, in the face of a similar rise in the serum phosphorus concentration and a fall in the serum calcium concentration. The increment of intact PTH to the change of ionized calcium was lower in the diabetic than in the control group (fig. 3).

Discussion

In the present study, we showed that in the diabetic patients the response of intact PTH to phosphate administration was lower than in nondiabetic subjects without renal failure. These results indicated that PTH secretion was reduced in the diabetic patients without renal failure as well as in the diabetic dialyzed patients. It is well known that there is a reduced incidence of hyperparathyroidism in diabetic dialyzed patients [1–3]. Morii et al. [4] reported fewer bone changes in dialyzed diabetic patients than in dialyzed patients with chronic glomerulonephritis but without diabetes: this could be due to the lower concentration of the carboxyl-terminal region PTH in the blood. Recently, Kikunami et al. [5] demonstrated that secondary hyperparathyroidism in dialyzed diabetic patients developed slower than in dialyzed patients without diabetes mellitus. It was pointed

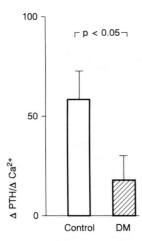

Fig. 3. The ratio between ΔPTH and ΔCa (the change from baseline 5 days after phosphate administration).

out that other metabolic hormones, such as growth hormone, C-peptide, insulin and renin may be impaired in diabetes mellitus [7]. Christlieb et al. [8] reported that the renin-aldosterone response to low sodium diet was subnormal in diabetic nephropathy. Studies of in vitro [9] and in vivo [10] systems showed direct stimulation of PTH release by β-adrenergic agonists. PTH secretion through the adrenergic nerve system may have been impaired in the diabetic patients with neuropathy. Therefore, microangiopathy should be regarded as a cause of the functional disorder of the parathyroid gland.

Active vitamin D metabolism is important as a modulating factor in the parathyroid hormone secretion. In diabetics, $1,25(OH)_2D$ is low [11, 12] or normal [13], so it is expected that parathyroid hormone secretion is affected. However, in this study the basal level was normal and the responsiveness to phosphate administration reduced. Moreover, in the young normal subjects [14], the $1,25(OH)_2D$ level did not change after phosphate administration, suggesting that the suppressive effect of phosphate on $1,25(OH)_2D$ production was effectively counterbalanced by the stimulatory effects of a rise in PTH. Therefore, the response of PTH may have been reduced because phosphate-induced suppression of $1,25(OH)_2D$ may not have occurred in the diabetic patients. However, in the study of Portale et al. [15], phosphate-induced suppression of $1,25(OH)_2D$ was observed only after a period of phosphate restriction. In these respects,

1,25($OH)_2$D may not play such an important role in PTH secretion in this study. Further study is needed including a sex-matched one.

Summary

To investigate the parathyroid function in diabetes mellitus, we performed an oral phosphate load in 6 diabetic patients and 6 nondiabetic subjects without renal failure (serum creatinine less than 1.5 mg/dl). Each subject received a total of 2.0 g of phosphate daily per os on 5 consecutive days. Blood and urine samples were obtained daily before and 2 h after the administration of phosphate in the morning. All subjects responded with a similar increase in the serum phosphorus concentration and fall in the ionized calcium concentration. Intact parathyroid hormone levels rose by 2.6-fold in the control subjects but by less than 1.5-fold in the diabetic subjects. It was concluded that hyporesponsiveness of the parathyroid hormone to phosphate administration was found in the diabetic patients without renal failure.

References

1   McNair P, Christensen MS, Madsbad S, Christiansen C, Transbøl I: Hypoparathyroidism in diabetes mellitus. Acta Endocrinol 1981;96:81–96.
2   Avram MM: Lower parathyroid hormone and creatinine in diabetic uremia. Contr Nephrol. Basel, Karger, 1980, vol 20, pp 4–8.
3   Avram MM, Lipner HI, Sadiquali R, Iancu M, Gan AC: Metabolic changes in diabetic uremia patients on hemodialysis. Trans Am Soc Artif Intern Organs 1976;22:431–438.
4   Morii H, Iba K, Nishizawa I, Okamoto T, Matsushita Y, Kikunami K, Inoue T, Inoue T: Abnormal calcium metabolism in hemodialyzed patients with diabetic nephropathy. Nephron 1984;38:22–25.
5   Kikunami K, Nishizawa Y, Tabata T, Nakatsuka K, Matsushita Y, Inoue T, Miki T, Morii H: Changes in parathyroid hormone in diabetic patients on long-term hemodialysis. Nephron 1990;54:318–321.
6   Auwerx J, Dequeker J, Bouillon R, Geusens P, Nijs J: Mineral metabolism and bone mass at peripheral and axial skeleton in diabetes mellitus. Diabetes 1988;37:8–12.
7   Hilsted J: Pathophysiology in diabetic autonomic neuropathy: Cardiovascular, hormonal and metabolic studies. Diabetes 1982;31:730–737.
8   Christlieb AR, Kaldny A, D'Elia JA, Williams GH: Aldosterone responsiveness in patients with diabetes mellitus. Diabetes 1978;27:732–737.
9   Sherwood LM, Hanley DA, Takatsuki K, Birnbaumer ME, Schneider AB, Wells SA Jr: Regulation of parathyroid hormone secretion; in Coop, Talmage (eds): Endocrinology of Calcium Metabolism. Proc. VIth Parathyroid Conf. Amsterdam, Excerpta Medica, 1978, pp 301–307.
10 Fischer JA, Blum JA, Binswanger U: Acute parathyroid hormone response to epinephrine in vivo. J Clin Invest 1973;52:2434–2440.
11 Frazer TE, White NH, Hough S, Santiago JV, McGee BR, Bryce G, Mallon J, Avioli LV: Alternations in circulating vitamin D metabolites in the young insulin-dependent diabetic. J Clin Endocrinol Metab 1981;53:1154–1159.
12 Imura H, Seino Y, Ishida H: Osteopenia and circulating levels of vitamin D metabolites in diabetes mellitus. J Nutr Sci Vitaminol 1985;31:27–32.

13   Heath H, Lambert PW, Service FJ, Arnaud SB: Calcium homeostasis in diabetes
     mellitus. J Clin Endocrinol Metab 1979;49:462–466.
14   Silverberg SJ, Shane E, Clemes TL, Bilezikian JP: Effect of oral phosphate administra-
     tion on major indices of skeletal metabolism in normal subjects. J Bone Miner Res
     1986;1:383–388.
15   Portale AA, Halloran BP, Murphy MM, Morris RC Jr: Oral intake of phosphorus can
     determine the serum concentration of 1,25-dihydroxyvitamin D by determining its
     production rate in humans. J Clin Invest 1986;77:7–12.

Takahiko Kawagishi, MD, Second Department of Internal Medicine,
Osaka City University Hospital, Abeno, Osaka (Japan)

Morii H (ed): Calcium-Regulating Hormones. I. Role in Disease and Aging.
Contrib Nephrol. Basel, Karger, 1991, vol 90, pp 223–227

Decreased Bone Mineral Density in Diabetic Patients on Hemodialysis

Hiroshi Nishitani[a], *Takami Miki*[a], *Hirotoshi Morii*[a],
Yoshiki Nishizawa[a], *Eiji Ishimura*[a], *Satoru Hagiwara*[a],
Kiyoshi Nakatsuka[a], *Makoto Yamakawa*[b]

[a]Second Department of Internal Medicine, Osaka City University Medical School,
Osaka, Japan; [b]Kidney Center, Shirasagi Hospital, Osaka, Japan

Metabolic bone disease leading to bone mineral loss and clinical symptoms is a well-recognized problem in patients with chronic renal failure [1]. Osteopenia and low levels of bone formation have been observed in diabetic patients with and without chronic renal failure [2, 3].

Renal osteodystrophy in diabetic patients on hemodialysis (DM-HD) might be somewhat different from that in nondiabetic patients on hemodialysis (non-DM,HD).

In this study, bone mineral density (BMD) was measured in patients with and without diabetes mellitus on hemodialysis, and factors that contribute to differences in calcium metabolism were identified.

Methods

Two studies were done. One was a comparison of BMD in non-DM,HD patients and DM-HD patients. The second was a comparison of possible factors affecting calcium metabolism in the higher and lower BMD groups in the DM-HD patients. In the first study, 60 patients on hemodialysis, 30 with and 30 without diabetes mellitus, were studied. All patients were men. In the second study, a total of 41 DM-HD patients, 26 men and 15 women, were studied.

BMD was measured by dual-energy X-ray absorptiometry (DEXA; Hologic QDR 1,000/W) in the third lumbar vertebra (L_3), head, pelvis, and whole body. Before the measurements of BMD, blood samples were drawn, and serum values (reference values in parentheses) of total calcium (4.2–5.1 mEq/l) and phosphate (2.5–4.5 mg/dl) were measured on an auto-analyzer (Hitachi 726). Serum immunoreactive parathyroid hormone (c-PTH) was measured by a c-terminal radioimmunoassay (Immuno Nuclear Corp., Minn.). The serum aluminum level was measured by flameless atomic absorption spectrophotometry.

Student's t test was used for statistical analysis.

Table 1. Comparison of BMD between the non-DM,HD and the DM-HD groups

	non-DM,HD	DM-HD
n	30	30
Age, years	58.7 ± 7.4	58.3 ± 7.8
Time since first dialysis, months	71.9 ± 36.6	$27.7 \pm 23.4***$
L_3 BMD, g/cm^2	0.954 ± 0.176	0.947 ± 0.152
Head BMD, g/cm^2	1.787 ± 0.233	$1.626 \pm 0.316*$
Pelvic BMD, g/cm^2	0.771 ± 0.037	0.760 ± 0.152
Whole-body BMD, g/cm^2	0.967 ± 0.093	0.874 ± 0.113
Serum		
Ca, mEq/l	4.90 ± 0.34	4.71 ± 0.38
P, mg/dl	6.17 ± 1.16	6.07 ± 1.65
cPTH, ng/ml	7.75 ± 7.32	$3.39 \pm 3.73**$
Dose of D_3, μg	0.38 ± 0.30	$0.20 \pm 0.21***$

$*p < 0.05$; $**p < 0.01$; $***p < 0.005$.

Results

Clinical and biochemical results and BMD in the first study are given in table 1. The patients were 58.7 ± 7.4 (mean $\pm$ SD) years old in the non-DM,HD group and 58.3 ± 7.8 years old in the DM-HD group. The difference was not significant. The time since the first dialysis was 71.9 ± 36.6 months in the non-DM,HD group and 24.7 ± 23.4 months in the DM-HD group; this difference was significant ($p < 0.005$). The L_3 BMD of the non-DM,HD group was 0.954 ± 0.176 g/cm^2 and that of the DM-HD group was 0.947 ± 0.152 g/cm^2. The BMD of the head for the non-DM,HD group was 1.787 ± 0.233 g/cm^2 and 1.626 ± 0.316 in the DM-HD group. The pelvic BMD was 0.771 ± 0.037 g/cm^2 in the non-DM,HD group and 0.760 ± 0.152 g/cm^2 in the DM-HD group. The whole-body BMD of the non-DM,HD group was 0.967 ± 0.093 g/cm^2 and that of the DM-HD group was 0.874 ± 0.113 g/cm^2.

The BMDs of the DM-HD group were lower in these areas and whole-body than those in the non-DM,HD group; the differences in the L_3 and in the pelvic BMD were slight, but a significant difference was found in the head ($p < 0.05$). The whole-body BMD of the DM-HD group was about 90% that of the non-DM,HD group, but the difference was not significant.

In both groups, the serum total calcium level was being controlled well at a mean of 4.90 ± 0.34 mEq/l in the non-DM,HD group and 4.71 ± 0.38 mEq/l in the DM-HD group. The serum phosphate level was

6.17 ± 1.16 mg/dl in the non-DM,HD group and 6.07 ± 1.65 mg/dl in the DM-HD group. There were no significant differences in the serum total calcium and phosphate values between the two groups.

The serum c-PTH level of the non-DM,HD group was 7.75 ± 7.32 ng/ml and that of the DM-HD group was 3.39 ± 3.73 ng/ml. The serum c-PTH level of the non-DM,HD group was significantly higher (p < 0.01).

The dose of 1α-hydroxy-vitamin D_3 (1α-OH-D_3) given to the patients for maintenance of the serum calcium level and suppression of secondary hyperparathyroidism was 0.38 ± 0.30 μg/day in the non-DM,HD group and 0.20 ± 0.21 μg/day in the DM-HD group. The daily dose of 1α-OH-D_3 taken by the non-DM,HD group was larger.

In the second study, factors which may contribute to the differences in BMD were compared in the DM-HD patients divided into higher and lower BMD of the head. The group with higher head BMD had a value 110% of the mean value or more; there were 20 such patients, 11 men and 9 women. The group with lower BMD of the head contained 21 subjects, 15 men and 6 women. The mean ages of the group with higher and the group with lower values were 63.2 ± 8.3 and 57.0 ± 9.9 years. Respectively, the time since the first dialysis was 27.8 ± 24.0 months in the group with higher values and 34.7 ± 34.6 months in that with lower values. The mean body weight of the group with higher BMD was 59.0 ± 12.1 kg and that in the group with lower BMD 52.9 ± 8.8 kg. The degree of obesity of the patients in the group with higher BMD was 112.1 ± 12.1% and that in the group with lower BMD 97.0 ± 14.3%; this difference was significant (p < 0.005). The mean height of the patients was not different in the two groups, at 158.2 ± 8.1 cm in the group with higher BMD and 160.8 ± 7.3 cm in the group with lower BMD.

Biochemical test results of the group with higher BMD and that with lower BMD were: serum total calcium, 4.88 ± 0.4 and 4.67 ± 0.35 mEq/l; serum phosphate, 6.35 ± 1.63 and 5.67 ± 1.63 mg/dl; serum aluminum, 2.20 ± 1.05 and 1.97 ± 1.36 g/dl; serum c-PTH level, 3.43 ± 2.92 and 4.50 ± 5.85 ng/ml. There were no significant differences in the biochemical test results, respectively, between the two groups. The dose of 1α-OH-D_3 in the group with higher values was 0.23 ± 0.22 μg/day and that in the group with lower values was 0.17 ± 0.10 μg/day. The difference was not significant (table 2).

Discussion

Important causes of bone mineral loss in patients on hemodialysis are secondary hyperparathyroidism, low plasma levels of $1.25(OH)_2D_3$, and

Table 2. Comparison of factors in calcium metabolism of DM-HD patients with high and low head BMD

	> 110%	< 110%
n (M/F)	20 (11/9)	21 (15/6)
Age, years	63.2 ± 8.3	57.0 ± 9.9
Time since first dialysis, months	27.8 ± 24.0	34.7 ± 34.6
Body weight, kg	59.0 ± 12.1	52.9 ± 8.8
Obesity, %	112.1 ± 12.1	97.0 ± 14.3*
Body height, cm	158.2 ± 8.1	160.8 ± 7.3*
Serum		
Ca, mEq/l	4.88 ± 0.40	4.67 ± 0.35
P, mEq/l	6.35 ± 1.63	5.67 ± 1.63
Al, μg/dl	2.20 ± 1.05	1.97 ± 1.36
cPTH, mEq/l	3.43 ± 2.92	4.50 ± 5.85
Dose of D_3, μg	0.23 ± 0.22	0.17 ± 0.10

100% = Mean BMD. *p < 0.005.

osteoporosis in women [1]. The BMD of men did not decrease with age, unlike the changes observed in women [4].

In diabetic patients, osteopenia has been reported [2]. In uremic diabetic patients, a decreased incidence of secondary hyperparathyroidism has been found compared with uremic nondiabetic patients [5]. Recently, the number of diabetic uremic patients on hemodialysis is rapidly increasing. More needs to be known about metabolic bone disease in diabetic uremic patients on hemodialysis.

In this study, significant differences were found in the head BMD, time since first dialysis, serum c-PTH level, and dose of 1α-OH-D_3 between the non-DM,HD group and the DM-HD group. The head BMD was lower, the time since first dialysis was shorter, the serum c-PTH level was lower, and the dose of 1α-OH-D_3 was less in the DM-HD group.

In a study reported elsewhere, we have found that whole-body BMD is negatively correlated with the serum c-PTH level and it is not correlated with the time since first dialysis. Differences in the time since first dialysis and in the serum c-PTH level seem not to be causes of the lower BMD in diabetic patients. Bone mineral loss was affected by vitamin D administration, and a lower dose of 1α-OH-D_3 may be the reason for the lower BMD in our DM-HD group, but the difference was only 0.18 μg/day. Thus, the reason is not clear.

To study the factors that affect the BMD in diabetic uremic patients on hemodialysis, a second study was performed here. In that study, diabetic uremic patients on hemodialysis were divided into two groups depending on their head BMD. Clinical and biochemical results were compared between the groups with higher and lower head BMD. Only the degree of obesity was significantly different.

In the second study we did not identify the reason for bone mineral loss in diabetic uremic patients. Various factors may participate in causing the difference in calcium metabolism between DM-HD and non-DM,HD patients.

Summary

Renal osteodystrophy in hemodialyzed patients with DM-HD shows different features from that in non-DM,HD. Two studies were done. One was a comparison of BMD in 30 non-DM,HD patients and 30 DM-HD patients. The second was a comparison of possible factors affecting calcium metabolism in the higher and lower BMD groups (n = 20/21) in the DM-HD patients. BMD was measured by dual-energy X-ray absorptiometry (DEXA; Hologic QDR 1,000/W) in the third lumbar vertebra (L_3), head, pelvis, and whole body. The BMDs of the DM-HD group were lower in these areas and whole body than that in the non-DM,HD group. A significant difference was found in the head BMD ($p < 0.05$). In the second study, factors which may contribute to the differences in BMD were compared in the DM-HD patients divided into higher and lower BMD of the head. The group with higher head BMD had a value 110% of the mean value or more. Clinical and biochemical test results (age, the time since the first dialysis, body weight, the degree of obesity, height, serum calcium, serum phosphate, serum aluminum, serum c-PTH level and the dose of 1α-OH-D_3) were compared. The degree of obesity of the patients with higher BMD was significantly larger than that with lower BMD ($p < 0.005$).

References

1   Sherrard DJ: Renal osteodystrophy. Semin Nephrol 1986;6:56–67.
2   Hui SL, Epstein S, Tohston CC Jr: A prospective study of bone mass in patients with type 1 diabetes. J Clin Endocrinol Metab 1985;60:74–80.
3   Morii M, Iba K, Nishizawa Y, et al: Abnormal calcium metabolism in hemodialyzed patients with diabetic nephropathy. Nephron 1984;38:22–25.
4   Gubta S, Luna E, Belsky J, et al: Photon absorptiometry for non-invasive measurement of bone mineral content. Clin Nucl Med 1984;9:435–439.
5   Vincenti F, Hattner R, Amend WJ, et al: Decreased secondary hyperparathyroidism in diabetic patients receiving hemodialysis. JAMA 1981;245:930–933.

Hiroshi Nishitani, MD, Second Department of Internal Medicine,
Osaka City University Medical School, 1-5-7 Asahi-machi,
Abeno-ku, Osaka 545 (Japan)

Subject Index

Numbers in *italics* refer to volume number

This volume
R